An
Introduction
to
Human
Physiology

An Introduction to Human Physiology

D. F. Horrobin

MTP
Medical and Technical Publishing Co Ltd

Published by

MTP
Medical and Technical Publishing Co. Ltd.
P.O. Box 55, St. Leonard's House, St. Leonardgate,
Lancaster

ISBN No. 852 000 48 0

First published 1973

Printed and Bound in Great Britain by
Hindson Print Group Ltd, Newcastle upon Tyne

Contents

An Introduction to Human Physiology Author's Preface

In many fields of study it is difficult to understand the significance of the part before one understands the whole. Yet one cannot understand the whole without a prior understanding of the parts. The dilemma is one of the most difficult problems to be solved by the teacher and in no subject is it more important than in physiology. In physiology more than in most subjects the part serves the whole and the whole serves the parts in an extraordinarily intimately integrated manner.

After teaching physiology at all levels for more than a decade, I have come to the conclusion that most students at all levels who study physiology in most hospitals, colleges and universities go through their courses in a fog of confusion and partial understanding. They cannot appreciate the working of each system until they understand the working of all the systems. For the fortunate few enlightenment dawns towards the end of the course but for many full understanding never comes.

There are many first class large texts of physiology which give students all the information they require. But the size of these books means that it is impossible to sit down and to read them through in order to get a quick, overall understanding of what physiology is all about. Furthermore, these larger books often unjustifiably assume a considerable amount of basic scientific knowledge. The better students may be able to cope with this, but many will find themselves groping. It seemed to me therefore that there was a need for a short introductory textbook of physiology which assumed little prior knowledge and which concentrated on an intelligent, lucid and up-to-date account of the important principles.

I have therefore tried to write such a book. Its aim is to provide a modern introduction to physiology and biochemistry. Medical students and others who require an extensive knowledge of physiology will find it a useful outline map describing the areas which they must later cover in more detail. For those such as nurses, physiotherapists, physical training instructors, occupational therapists and others whose need for depth is not so great, it can stand on its own as a text which although clearly written and easy to understand offers a modern account of physiology in its relation to human biology and medical science.

D. F. HORROBIN.

1 Introduction

The world of living things is conveniently and conventionally divided into two great groups, the animals and the plants. Broadly speaking, the important feature which distinguishes plants is that they can manufacture most of the substances they require by trapping and using various forms of outside energy, in particular the energy of sunlight. In the process of photosynthesis they utilise the energy of light to build up complex chemical substances from relatively simple ones.

In contrast, animals lack the ability to use light or any other form of outside energy. Instead they must obtain the energy they require by breaking down complex substances which ultimately they always obtain from plants. Plant-eating animals such as cows and sheep obtain these substances directly. Carnivores obtain them indirectly after they have passed through the bodies of other animals.

During the past few decades, the properties of a third group of organisms have been determined. These are the viruses and they are particularly important in the causation of many human diseases such as smallpox, polio and measles. They are unusual in that while they can survive outside the living cells of animals or plants, they can only reproduce themselves while they are parasites actually inside plant or animal cells. They use the mechanisms which they find within the cells for the manufacture of more virus material.

THE CONCEPT OF THE CELL

All plants and animals are made up of very large numbers of sub-units known as cells. In both groups of living organisms the simplest species consist of single cells. We can learn a great deal about the functioning of such complex creatures as human beings by studying the structure and the requirements of a relatively simple single-celled animal such as the familiar amoeba.

All animal cells have certain structural features in common. The three most important of these are the cell membrane, the nucleus and the cytoplasm. The cytoplasm is the material within the cell membrane but outside the

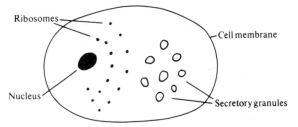

Figure 1.1 An outline sketch of a cell.

nucleus. The cell membrane is an extremely important structure about which as yet we know relatively little. It is not an inert barrier which lets virtually anything in or out of the cell. It is highly selective and exerts a very active control over substances which try to cross it. Some substances, like sodium, for example, are kept out of the cell and if they try to gain entry they are forcibly ejected. Potassium on the other hand is kept inside the cells. As a result the concentration of potassium inside cells is high and that outside cells is low. The reverse is true of sodium. We are only just beginning to understand the ways in which the cell membrane functions and full knowledge lies far in the future.

The other universally found structure is the nucleus. This contains the chromosomes which seem to consist mainly of deoxyribonucleic acid or DNA. The nucleus is essential for the normal development of the cell because the chromosomes contain the plans for the manufacture of all the substances and structures required within the cell. It is also involved in the day-to-day running of the cell's activities. A breakdown of this control exerted by the nucleus can probably account for many diseases which as yet are poorly understood, and in particular, for cancer.

THE CYTOPLASM

This is the part of the cell outside the cell nucleus but within the cell membrane. It varies considerably in its structure depending on the function of the cell. In most cells the cytoplasm contains a series of folded membranes on which are tiny granules known as the ribosomes. In the ribosomes the manufacture of the proteins which the cell needs takes place. In cells which secrete substances, such as cells which manufacture the digestive juices or the hormones, there are often other granules in the cytoplasm known as secretory granules. These contain the secretions wrapped up in what appear to be little packets. At the appropriate time the cell can push them out across the cell membrane, so releasing their contents. In most cells which secrete things there is a complex structure, the Golgi apparatus, which is thought to be the place where the secretory granules are manufactured.

CELLS WITH A 'BRUSH BORDER'

The brush border is a structure which is found in many of the cells which line the gut. It is found only on one side of each cell, the side which faces the gut cavity. When looked at with an ordinary microscope it looks as though

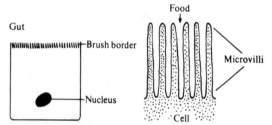

Figure 1.2 Cell from the lining of the gut showing the brush border. The right hand diagram shows the border greatly magnified.

the cell has a brush-like structure on that side hence the name 'brush border'. But when looked at under the much greater magnification of the electron microscope, it can be seen to consist of innumerable finger-like processes known as microvilli. The function of the microvilli is to increase the surface area of the cell. The microvilli thus greatly increase the surface area of the cell membranes facing the gut cavity. The reason for this is that all food must be absorbed from the gut by crossing the barrier of the surfaces of the lining cells. Each fragment of surface can only absorb food at a limited rate and so the rate of absorption can be greatly increased by increasing the amount of surface area available. In some diseases such as sprue and coeliac disease, the intestinal wall is damaged and many of the cells lose their microvilli. As a result food cannot be absorbed properly and passes straight through the gut into the faeces causing diarrhoea. Because relatively little of the food can be absorbed, the patient effectively becomes starved. The general term for this condition is the malabsorption syndrome.

THE REQUIREMENTS OF A SINGLE-CELLED ANIMAL

All single-celled animals live in water, in ponds, in rivers, in lakes or in the sea. It is essential to recognise that such single animal cells are very delicate structures which are

easily destroyed. In particular, they can be damaged by three types of changes in the fluid which bathes them.

1. Changes in temperature. They can be killed if the temperature of the water becomes either too high or too low.

2. Changes in the acidity or alkalinity of the fluid. Most of the water in which such animals live is nearly neutral, being neither very acid or very alkaline. If by chance the water does become acid or alkaline, the animals within it are often killed.

3. Changes in the chemical composition. Each animal tends to be adapted to a particular chemical composition. For example, animals which live in the sea are accustomed to a high salt content and are killed if they are placed in fresh water. Conversely most freshwater animals are killed if they are placed in the sea. Each animal is therefore adapted to a particular chemical composition of the surrounding fluid and can be killed if this changes very much.

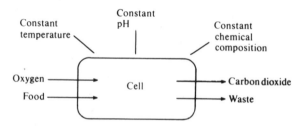

Figure 1.3 The requirements of a living cell.

Most single-celled animals do not have to worry about these problems. The temperature changes of the water in which they live are very much smaller than the temperature changes on land. The chemical composition and pH of most lakes and rivers, and of the sea are virtually constant. Single-celled animals do not therefore have to make any attempt to regulate the environment in which they live.

Apart from requiring constancy in certain environmental conditions, animal cells need to obtain certain things from their surroundings. The most important of these are food materials and oxygen. Food is obtained by engulfing tiny particles of animal or vegetable origin floating in the water. The process is known as phagocytosis and the food is taken into the cell across the cell membrane. Some of the cells in the human body are capable of phagocytosis. The most important are some of the white cells in the blood and some of the cells which line

Figure 1.4 The process of phagocytosis.

blood vessels in the liver, the spleen and the bone marrow. Their function is to remove unwanted particles of solid matter from the blood, ranging from bacteria to old and dying red blood corpuscles.

Oxygen is essential for the processes in which food is broken down to supply energy. It also is taken in directly from the surrounding water, this time by the mechanism known as diffusion. Diffusion is very important in all living creatures, including humans, and it is important to understand it. Diffusion occurs because all substances tend to move from regions where they are in high concentration to regions where they are in low concentration: the process continues until no concentration differences remain. This may very simply be illustrated by putting a drop of blue ink into a jar full of clear water. At the instant of putting the drop in, the concentration of the blue dye is very high in the drop and zero in the water. But as soon as the drop enters the water, even if the water is absolutely still and is not stirred, the dye begins to spread through the water from areas where there is a lot of dye (where it is high in concentration) to areas where there is little or no dye (where it is low in concentration). Eventually by the process of diffusion the dye becomes evenly distributed throughout the water and there are no concentration differences left.

The law of diffusion does not apply only to ink and water. It applies to all gases and to all liquids. All substances in gases and liquids tend to move from areas where they are high in concentration to areas where they are low in concentration. In the case of a single animal cell in a lake or in the sea, the cell uses up oxygen and therefore the concentration of oxygen within the cell falls below its concentration in the surrounding water. There is therefore a concentration difference between the oxygen level in the cell and the oxygen level in the water. Oxygen therefore diffuses from the water into the cell. Precisely the same thing happens in the human body. Cells use up oxygen and the oxygen concentration within them falls below the oxygen concentration in the blood. As a result oxygen moves from the blood into the cells.

EXCRETION OF SUBSTANCES

Single cells do not only take things from the water in which they live. They also pass things out into the surrounding water. The burning of food to give energy yields carbon dioxide and many other waste products such as sulphuric acid and ammonia. As these substances are produced, their concentration inside the cell rises above their concentration in the surrounding water. They therefore move out of the cells into the water by the process of diffusion. They do not usually significantly alter the composition of the water because the volume of that water is so vast compared to the volume of a single cell. The waste products therefore rapidly become diluted. Occasionally this is not true. For example, under certain conditions tiny animals known as dinoflagellates can multiply in the sea in vast numbers. They are reddish in colour and there may be so many of them that the sea as a whole appears to be red. They produce some highly poisonous waste products and these can destroy all other living things in the vicinity and make animals like shellfish too toxic for human beings to eat. However, this sort of thing is very unusual and single-celled organisms normally do not alter to any serious extent the composition of the water in which they live.

THE NEED FOR MOVEMENT

Plants can usually obtain all they need without moving. They utilise the energy of the sun and collect the very simple chemicals they require from the surounding water or from the soil. But animals have to move in search of food. Tiny single-celled animals go about this in two different ways. Some like the amoeba move by pushing out part of their cell membrane and then allowing the contents of the cell to flow into the part which has been pushed out. This is known as amoeboid movement and it is also important in the human body. Some of the white cells of the blood which are essential for the protection of the body against bacterial invasion move through the tissues in an amoeboid fashion and then engulf the bacteria by means of phagocytosis.

Some other single-celled animals are covered by masses of tiny hair-like structures known as cilia. These thrash the water in a coordinated way generating currents which push the water backwards and the animal forwards. Again, cilia are important in humans but not in quite the same way. Many human cells which line the internal surfaces of the body, such as the respiratory tract or the Fallopian tube which goes from the ovary to the uterus, have cilia on their surfaces. The cells are fixed in position and lashing of the cilia cannot move them. But the ciliary motion can move fluid lying on the surface of the cell. Thus the cilia in the respiratory tract generate currents which steadily wash mucus, dust, bacteria and many other types of material out from the lungs into the throat where they may be either coughed up as sputum or swallowed and destroyed by the secretions of the digestive tract. This process is vital for keeping the lungs clean. Some poisonous gases paralyse the cilia. As a result all the material cannot be removed from the lungs and the patient often dies quickly from pneumonia. In the Fallopian tubes, the cilia generate currents which steadily wash the secretions of the cells lining the tubes down from the ovaries into the uterus. This current of fluid takes with it the eggs and ensures that they arrive safely in the uterus.

REPRODUCTION

If living species are to survive, the individual members of that species must reproduce themselves. Single-celled organisms usually do this by a simple process of cell division but this is obviously impossible for larger organisms. Most larger organisms, including of course humans, reproduce themselves by some form of sexual activity.

THE FUNCTIONING OF LARGER ANIMALS

It is obvious that large animals consisting of millions of cells face many more problems than single-celled organisms. Even in large animals, however, each single cell is bathed in fluid, usually known as the extracellular or inter-stitial fluid. This fluid is not in free and easy communication with the surroundings of the animal. This section is concerned with the ways in which large animals, including human beings, cope with the difficulties posed by their size.

LARGE ANIMALS WHICH LIVE IN WATER

Living on land introduces even further complications, and so first of all we shall consider a large animal which lives in water such as a fish. Perhaps the most obvious feature of a large creature like this which distinguishes it from a single-celled animal is that there are millions upon millions of cells in the body and each cell does not carry out all the functions which the body as a whole needs. In the amoeba, the same single cell collects food, oxidises the food to supply energy, moves about, excretes waste products and is responsible for reproduction. In larger animals, each cell is specialised to carry out only some small aspect of the functioning of the body as a whole. Cells which are specialised to carry out a particular function tend to be gathered together in groups forming the various organs of the body.

The whole cell surface of the amoeba is in contact with the surrounding water. The cell can easily take from its surroundings what it needs and discharge its waste products directly into the lake or into the sea. In a fish, most of the cells lie deep within the body, far from the water in which the fish swims. Most of the cells in a fish cannot directly obtain food and oxygen from that water, nor can they discharge carbon dioxide, ammonia and other waste material into it. The cells in a fish, it is true, are surrounded by the water of the extracellular fluid: it is from this fluid that they must obtain their oxygen and food and into this fluid must pass their carbon dioxide and other waste.

THE CIRCULATORY SYSTEM

In both the fish and the amoeba, each cell is surrounded by water. The main difference is that the volume of water surrounding an amoeba in a lake, a river or the sea is vast in comparison to the size of the amoeba. It is therefore

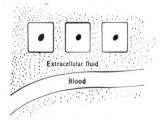

Figure 1.5 The relationship between the blood, cells and inter-stitial (extracellular) fluid.

impossible, except in unusual circumstances, for single-celled organisms to alter the composition of the water which surrounds them by what they take out of it or put into it. The situation in large animals is very different.

The volume of the extracellular fluid surrounding each cell is very small. In most cases the volume of this fluid is much smaller than the volume of fluid actually within the cell. A living cell can therefore very rapidly exhaust all the oxygen and food in the fluid immediately surrounding it. Such a cell can also rapidly alter the acidity and chemical composition of the surrounding fluid. Most cells produce quite a lot of heat and the temperature of the fluid can therefore rise. In fact any living cell, if it had to make do solely with the extracellular fluid immediately surrounding it, would very quickly kill itself by using up the oxygen and poisoning itself with waste products. This is why it is so essential to have a circulatory system which can bring oxygen and food to the extracellular fluid and can take away carbon dioxide, heat and other waste products. The way in which the circulation works can be illustrated with reference to oxygen and carbon dioxide. A living cell continually uses oxygen and this lowers the concentration of oxygen inside it below that of the extracellular fluid. Oxygen therefore flows by diffusion from the extracellular fluid into the cell. Because the

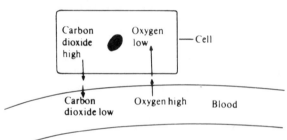

Figure 1.6 The movements of oxygen into and carbon dioxide out of cells.

volume of the extracellular fluid is so small it, too, is rapidly depleted of oxygen and its oxygen concentration falls well below the oxygen concentration of the blood. Oxygen therefore diffuses from the blood into the extracellular fluid. The cells are supplied with oxygen because of the process of diffusion. Similarly, living cells continually produce carbon dioxide. This raises the carbon dioxide concentration of the cell above that of the extracellular fluid and carbon dioxide therefore diffuses out of the cell into the fluid. In turn the carbon dioxide concentration in the extracellular fluid rises above that in the blood and so the carbon dioxide passes into the blood which carries it away.

It is clear that if the circulatory system is to work effectively it must bring the blood into close contact with the extracellular fluid surrounding every single cell in the body. It does this by an extremely complex system of blood vessels, the smallest of which are known as capillaries and have walls only one cell thick. There are capillaries in the immediate vicinity of every cell and it is across their walls that the exchange between the blood and extracellular fluid takes place. The blood is of course continually pumped around the circulatory system by means of the heart.

THE CONSTANCY OF THE INTERNAL ENVIRONMENT

If cells in a large animal are to survive, it is essential that the temperature and composition of the fluid immediately surrounding them must be kept constant. In the words of a famous French physiologist, Claud Bernard, the constancy of the internal environment of the cells is the main problem facing animal life. As we saw in the previous section, this constancy is maintained by the use of the blood which brings oxygen and food to the extracellular fluid and takes carbon dioxide, heat and other waste substances away. The maintenance of the constancy is often known as homeostasis.

The blood itself has only about one eighth of the volume of all the cells in the body and so, like the fluid surrounding the cells, it too is in danger of being rapidly depleted of its oxygen and food materials and of having its acidity, temperature and chemical composition drastically altered by the waste products of the cells. In fact, of course, the temperature and chemical composition of the blood remain remarkably constant. This is because the blood itself is served by a number of organs whose function is to maintain the constancy of its temperature and composition. The important organs which do this in a creature like a fish which lives in water are the following:

1. *The Alimentary Tract or Gut*

Food is taken in at the mouth and passed along the alimentary canal where it is first broken down into tiny particles (digested) and then taken into the blood across the gut wall (absorbed). This ensures that the blood receives a supply of food material.

2. *The Liver*

The supply of food is usually intermittent, and if the blood relied entirely on the food absorbed from the gut, there would be long periods when no food was being taken in when the concentration of food materials in the blood would be in danger of becoming very low. This situation is partially counteracted by the liver and some other organs which store food when it is plentiful and is being rapidly absorbed from the gut and which then release some of their stores when the gut is empty and no food is entering the blood that way.

3. *The Respiratory System*

In fish this consists of the gills, and in land animals of the lungs. In the gills the blood comes into close contact with a continual stream of water drawn from the surrounding environment and which has a relatively high level of oxygen and a low level of carbon dioxide. Oxygen therefore diffuses in and carbon dioxide diffuses out, so keeping the oxygen and carbon dioxide levels of the blood constant. The lungs are continually supplied with large amounts of fresh air in the process of breathing. Air has a high oxygen content and a very low carbon dioxide content, and so land animals too can obtain oxygen from and lose carbon dioxide to their surrounding environment.

4. *The Kidneys*

These paired organs are responsible for the excretion of acid, ammonia and other waste materials. In fish these substances can be simply poured into the surrounding water but in land animals they must be stored in the bladder which can then be emptied at intervals.

SPECIAL PROBLEMS OF MAN AND OTHER LAND ANIMALS

Animals which live in water, even those which are very large, have three important advantages over those which, like man, live on land. The land animals must find ways of solving these three major problems.

1. Water-dwelling animals have an unlimited supply of water, whereas land animals obviously do not. A previously healthy man will die within 2 or 3 days if he is totally deprived of water. If he has water, but is totally deprived of food, he can survive for several weeks. Water is therefore of much more importance than food, and land animals must always ensure that they have adequate supplies for drinking.

2. Water-dwelling animals can excrete their toxic waste materials directly and immediately into the surrounding water, which then dilutes the waste matter and renders it harmless. Land animals cannot do this. They must excrete their waste in a relatively small volume of urine and they must store their urine for periods before it is excreted. They cannot therefore afford to excrete highly toxic materials directly. Instead land animals must quickly detoxify such materials and thus render them harmless. This particularly applies to ammonia, a highly dangerous product of protein metabolism. In land animals this must rapidly be converted by the liver to the relatively harmless substance urea. Only then can it safely be excreted in the urine.

3. Temperature regulation. The rates of the chemical reactions on which the working of the whole body depends are very much influenced by temperature. When the temperature falls, the reactions become slower and slower and an animal becomes more and more sluggish. When the temperature rises, the reactions go faster and faster and if they become too rapid the animal may die. Animals which live in water and particularly those which live in the sea are largely protected from violent temperature changes since the temperature of water never rises so high or falls so low as that of the land. Land animals face much bigger problems. Those whose body temperature depends on the environment, the so-called poikilotherms or cold-blooded animals, cannot function at all in winter because their body temperature never rises high enough. They therefore remain inert for half the year. Two groups of animals, the mammals to which man belongs and the birds, have solved this problem by making their body temperature independent of that of the environment. They are the so-called homoiotherms or warm-blooded animals which maintain a constant blood temperature, in the case

of man in the region of 37 °C or 98 °F. This constancy can be achieved only by having complex mechanisms which warm the body when it is too cold and which cool it when it is too hot. These mechanisms enable the warm-blooded animals to function effectively all the year round.

SUPPORT AND MOVEMENT OF LARGE ANIMALS

Large animals, like man, must have a system of supporting the body so that it does not collapse under its own weight. This is achieved by the possession of a body skeleton to which all the soft tissues of the body are ultimately attached. But it is also essential to be able to move the body easily and so it is no use having an absolutely rigid central support. Instead the skeleton must have many individual bones connected together by flexible joints. In order to move the joints, there must be energy generating structures—the muscles—which pull on the bones and thus move the body in appropriate ways. It is possible to ensure that movements are appropriate only by controlling them effectively. This is the job of the central nervous system (CNS). The sensory nerves constantly supply it with a barrage of information about the position of the body and the state of contraction of the muscles. The CNS then sends instructions out along the motor nerves. In this way the movements of the body are made appropriate to its needs.

2 The Maintenance of a Constant Internal Environment

All the systems of the body, if they are to function effectively, must be subjected to some form of control. Food intake must be controlled by employing the device of appetite, otherwise too much or too little food would be eaten. The breathing rate must be controlled so that oxygen can be supplied and carbon dioxide removed at precisely the right rate. The working of the kidney must be controlled so that water and waste materials are removed neither too slowly nor too quickly.

The precise control of body function is brought about by means of the operation of the nervous system and of the hormonal or endocrine system. The most important structures in the control system as a whole are the brain and the spinal cord, often together known as the central nervous system or CNS.

The most important thing to note about any control system is that before it can control anything it must be supplied with information. Before the CNS can decide whether or not a person should be hungry, it must be supplied with information about the level of food materials present in the blood. Before it can decide whether a person should breathe more slowly or more quickly it must know how much oxygen and carbon dioxide are present in the blood. The same is true for the control of the kidney and of the activity of all other organs in the body.

Therefore the first essential in any control system is an adequate system of collecting information about the state of the body. This is done by means of a complex system of sensory receptors whose function is to sense the state of all aspects of bodily function. There are receptors for blood glucose, for oxygen, for carbon dioxide, for the body water content, for blood acidity, for body temperature and for a myriad other things. Some of these receptors lie within the brain itself. Some are at the ends of long nerve fibres known as sensory nerves which transmit the information they collect back to the CNS in the form of nerve impulses. In this way the CNS can build up a comprehensive picture of what is happening all over the body.

Once the CNS knows what is happening, it must then have a means for rectifying the situation if something is going wrong. There are two available methods for doing this, by using nerve fibres and by using hormones. The motor nerve fibres are long, thin fibres which carry instructions from the CNS to the muscles and glands

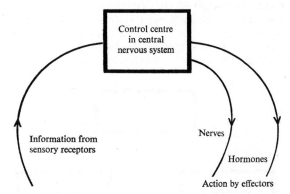

Figure 2.1 The basic components of most biological control systems.

throughout the body. For example, if breathing is too slow so that the oxygen content of the blood is falling and the carbon dioxide content is rising, then the motor nerves can tell the muscles of the diaphragm and chest wall to move more rapidly and more deeply so returning the situation to normal.

The other way that the CNS can act is via hormones. Hormones are chemicals which are released by glands known as endocrine glands. The hormones are secreted into the blood and are carried by the circulatory system all over the body. As far as hormones are concerned the brain acts via the pituitary gland, a minute structure lying in an inaccessible position in the centre of the skull at the base of the brain. The pituitary secretes a large number of hormones which are carried by the blood to every part of the body. The rate of secretion of each one of these is under the direct control of the brain. Some of the pituitary hormones act directly in their own right: growth hormone which helps to control growth and the level of blood glucose, and antidiuretic hormone (ADH) which governs the amount of water put out by the kidneys are examples of this type. Other pituitary hormones, however, are known as trophic hormones because they exert their effects by altering the behaviour of other endocrine glands throughout the body. For example, thyroid-stimulating hormone (TSH) from the pituitary, controls the behaviour of the thyroid gland which in turn governs the rate of working of many chemical reactions. Adrenal corticotrophic hormone (ACTH) from the pituitary, controls the behaviour of the adrenal cortex which in turn itself controls many different aspects of the function of the body.

By using the nervous and endocrine systems, the body is therefore able to control the behaviour of all the organs which help to maintain the internal environment of the body constant. Only if this constancy is maintained can the cells survive. If the regulating mechanisms fail, the fluids become poisoned and the cells and the organism as a whole die.

This chapter describes the principles on which the body's homeostatic mechanisms work.

BASIC PRINCIPLES OF CONTROL

The fundamental components which a control system of any type must possess can be clearly seen in the design of a simple household thermostat which controls the temperature of a room in a cold climate. The thermostat has

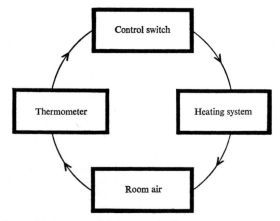

Figure 2.2 The essential elements of a thermostat.

three essential components, a thermometer, a source of heat, and a switch for turning the heat source on and off. The thermometer is linked to the switch and the two are so designed that whenever the temperature of the room falls below a certain level, the heat source is switched on. Whenever the temperature rises above that level the heating is switched off. The cycle of events is repeated again and again as the temperature of the room oscillates

around its set level, being pushed up by the heating system and pulled down by the cold outside environment. This oscillation about a fixed point is sometimes known as 'hunting'.

The magnitude of the swings in the temperature of the room depends on two things, the sensitivity of the thermometer and switching system and the delicacy of the control of the heat source. If the thermometer turns on the switch when the temperature falls only a fraction of a degree below the set level, X, and if the heating is turned up only gently so that the temperature does not rise very rapidly and overshoot X, the magnitude of the oscillations will be small. On the other hand, if the thermometer operates the switch only when the temperature has fallen several degrees below X and if the heat source is very powerful and poorly regulated, the temperature, while remaining centred on X may oscillate over a wide range.

Most thermostats have at least one other component and that is a device whereby the set level may be altered. There is usually a knob which can modify the behaviour of the switch so that instead of maintaining the temperature close to X, it maintains it close to Y. The heat source is turned on and off at a room temperature in the vicinity of Y instead of one in the vicinity of X.

DISTURBANCE DETECTORS

A point worth noting about the thermostat and one which applies to most physical and biological control systems is the siting of the sensing device, in this case the thermometer. The device is not placed at a point where it can measure the intensity of the forces tending to disturb the system. If the thermometer were to do that, it would have to be placed on the outside wall where it could measure the temperature of the outside environment and thus the intensity of the cold with which the heating system had to cope. In practice, sensing devices which measure the intensity of the forces tending to disturb a system have been found inadequate for accurate control. In the remainder of this book such devices will be called disturbance detectors. As an experiment, the thermometer of a thermostat can be put outside the house whose internal temperature is to be controlled. It can be linked to the heating system so that the colder the outside environment becomes, the more heating is turned on. If such a system is tried it is found that the temperature inside the house fluctuates wildly and the result is most unsatisfactory. The system fails primarily because there is no simple direct relationship between the environmental temperature and the temperature inside the house.

In any real house there is constant interference from uncontrollable variations in the rate of heat production and heat loss. People produce heat and the more people there are in the house, the greater will be the heat production. A thermometer on the outside wall could not cater for this. People also open and close curtains and windows and doors thus altering the rate of heat loss. Again a thermometer on the outside wall which simply measures the environmental temperature cannot take these factors, which have important effects on the internal temperature, into consideration. The control system is therefore certain to fail.

MISALIGNMENT DETECTORS

This means that in any good control system the sensing device should not measure the magnitude of a particular disturbing force. Instead it should measure the result of the interaction between the stress which is tending to disturb the status quo and the factors which are tending to maintain the steady state. This means that it should monitor the state of the factor whose constancy is desired. Therefore in a good thermostat the thermometer is put inside the room whose temperature is to be kept constant and not on the outside wall of that room. The temperature inside the room is the resultant of all the factors influencing heat production in the room and the heat loss from it. The magnitude and behaviour of each individual factor do not need to be understood for the system to work. It is sufficient for the thermometer simply to measure the temperature of the room which is the result of the interaction of all the various factors. The thermometer reading depends on the balance between all the factors causing heat production and all the ones promoting heat loss. Ideally at the desired temperature, heat loss just balances heat production. If heat production rises this is detected by the thermometer and the heating is turned down. If heat loss increases, the temperature falls, this is again detected by the thermometer and the heating is turned up. Such sensing devices in control systems are therefore sometimes known as misalignment detectors, because they detect any deviation from the desired value which may result from an imbalance of the forces acting upon the factor which is being controlled.

Although it may seem superfluous to say so, it is essential to remember that in a good control system the factor which is being controlled is the one which remains constant in value. If the system is disturbed in any way, it is the factor which counteracts the disturbance which fluctuates. The level of the factor which is being controlled should not alter. In the thermostat example, if the system is a good one, the reading of the thermometer will remain approximately constant no matter what the cold stress may be. It is the output of the heating system which will vary in order to cope with the varying intensity of the cold stress. But although the heat output varies while the thermometer reading remains constant, this does not mean that there is no connection between the two. Some people so misunderstand control systems that they believe that if the heating part of a thermostatic control device is functioning at a very high level, then the thermometer reading must also be way off the desired mark in order to provide the necessary intense stimulus for such a big response. This very elementary point is important and has so often been thoroughly misunderstood by biologists that it is worth describing in a very simple way the sequence of events which occurs in a household thermostat when the temperature of the outside air steadily becomes colder.

1. The cold stress exceeds the existing heat output and the temperature of the room falls slightly.

2. The fall in temperature is detected by the thermometer and in order to restore the balance between heat loss and heat production the power of the heating system is turned up by the control mechanism.

3. The temperature of the room returns to the desired level.

4. If the cold stress continues to increase, again the balance between heat loss and heat production will be upset. The thermometer reading will fall, heat output will be further increased and the temperature will again get back to its set level.

5. If the outside temperature falls further, the cycle may be repeated many times. Eventually the heating system will be roaring full blast but the room temperature will remain what it was before the cold stress began. This situation is a consequence of the fact that the thermometer is acting as a misalignment detector and not a disturbance detector. It simply measures the result of the interaction between heat loss and heat production. As the environmental temperature falls, the heating system is turned up correspondingly.

MISALIGNMENT DETECTORS AND RESPIRATION

There are several instances in biology where research workers have failed to understand that a massive effector response does not have to be produced by a large deviation from the normal level of the factor which is supposed to be kept constant. A classic example was a violent controversy not so many years ago about the mechanism by which the rate of ventilation (breathing) is increased in mild to moderate exercise. It had long been known that there are sensitive receptors for measuring the concentration of carbon dioxide in arterial blood. If the arterial carbon dioxide level rises the information is sent to the brain where the control centre orders an increase in the ventilation rate. The faster rate of breathing then gets rid of the carbon dioxide more rapidly. Under normal circumstances the receptors act as misalignment detectors: carbon dioxide enters the blood in the capillaries and the carbon dioxide-rich blood passes via the veins to the heart. But only after the blood has been passed through the lungs where the carbon dioxide is normally removed is the carbon dioxide level measured. The receptors measure the result of the interaction between the action of the peripheral organs in adding carbon dioxide to the blood and the action of the lungs in removing it. Only if there is an imbalance is the ventilation rate changed.

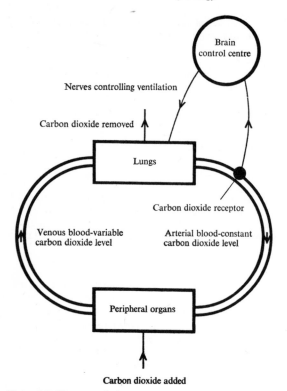

Figure 2.3 The control of ventilation rate.

The increase in ventilation rate in exercise can therefore be relatively simply explained as follows (this is an oversimplified but nevertheless reasonably valid interpretation):

1. The exercising muscles produce large amounts of carbon dioxide which raise the carbon dioxide content of the venous blood above normal.

2. At the resting rate of ventilation, the excess of carbon dioxide is not all removed during the passage through the lungs and so the carbon dioxide in the arterial blood rises.

3. This increase is detected by the arterial receptors. The information is sent to the brain and the control centres in the brain increase the ventilation rate in order to cope with the increased rate of carbon dioxide production. The arterial carbon dioxide level returns to normal. Nevertheless it is the arterial carbon dioxide receptors which have brought about the increased ventilation rate, even though as the result of the operation of the control system the arterial carbon dioxide level is held approximately constant.

But some physiologists failed to see this. They reasoned that since the arterial carbon dioxide level does not rise in moderate exercise, the increase in ventilation could not be brought about by the functioning of the arterial carbon dioxide receptors. Many elaborate experiments were carried out in the effort to find other factors which could stimulate ventilation. Only recently has it been understood that the fact that the arterial carbon dioxide level remains constant is the very best argument for the importance of carbon dioxide in controlling ventilation in exercise. In contrast, if the arterial carbon dioxide level were to fluctuate considerably one might then suspect that the changes in ventilation were unrelated to the arterial carbon dioxide level. Then one might start looking seriously for other factors.

NEGATIVE AND POSITIVE FEEDBACK
A feedback mechanism is said to exist when a change in a system leads to a sequence of events whose end result is to produce another change in the same system. That may sound complex but a few examples will make it clear.

There are two main types of feedback, negative and positive. The former is by far the most common in biological situations. With negative feedback, if a system is disturbed, that disturbance sets in motion a train of events which counteract the disturbance and tends to restore the system to its original state. The negative feedback principle has therefore been called the law of sheer cussedness because the mechanism resists any attempts to change the system: negative feedback mechanisms therefore tend to act in favour of stability. As might be expected from its name, positive feedback is quite different. In this case a disturbance in a system sets in motion a train of events which increases the disturbance still further. Positive feedback mechanisms therefore lead to instability and are rare in biology.

NEGATIVE FEEDBACK
The thermostat is an excellent non-biological example of a negative feedback mechanism. If the temperature falls, the change is detected by the thermometer and a sequence of events is set in motion to return the temperature to its set value. Because the end result of the operation of a negative feedback mechanism is stability, it is not surprising that there are innumerable biological examples of such mechanisms. The body's own temperature control system is one. If the body temperature rises, the change is detected by receptors which sample the arterial blood. The information is sent to the control system in the brain and this sets in motion mechanisms aimed at increasing the rate of heat loss. The rate of blood flow to the skin is greatly increased, so carrying heat from the deep organs to the surface where it can be lost to the surrounding environment. Sweat is secreted on to the skin surface and as it evaporates the skin is cooled. On the other hand if the body temperature falls, the rate of skin blood flow is reduced and shivering may be started in order to increase heat production. In both cases a change in body temperature leads to events which reverse that change and restore the system to normal.

The control of the carbon dioxide level of arterial blood is another good example which has already been discussed. If the carbon dioxide level rises as the result of exercise, this is detected by the arterial receptors. The information is conveyed to the brain and the ventilation rate is increased to get rid of the carbon dioxide excess. If the carbon dioxide level falls, then the rate of ventilation also falls and carbon dioxide accumulates until it reaches its normal level.

The output of hormones from most of the endocrine glands is also controlled by negative feedback mechanisms. The adrenal cortex will serve as an example. One of

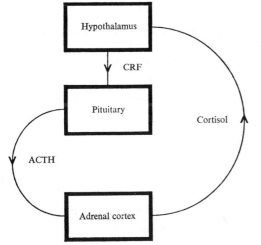

Figure 2.4 The negative feedback mechanism which controls the level of blood cortisol.

the most important hormones which it produces is known as cortisol (hydrocortisone). The rate of production of cortisol is determined by the blood level of adrenocorticotrophic hormone (ACTH) which is secreted by the anterior pituitary gland at the base of the brain. The anterior pituitary has no nervous connections with the part of brain immediately above it (the hypothalamus) but it does have an unusual blood supply. Blood which has been to the hypothalamus travels down to the

anterior pituitary by means of a special system of vessels: substances secreted into this blood by the hypothalamus therefore go directly to the anterior pituitary. We now know that the brain controls the anterior pituitary by secreting into these pituitary vessels releasing factors which govern the secretion of the pituitary hormones. The output of ACTH from the pituitary is governed by the secretion of corticotrophin releasing factor (CRF) from the hypothalamus. When the output of CRF rises, so does the output of ACTH: the ACTH in turn then stimulates the adrenal cortex to secrete cortisol. But by means of a negative feedback mechanism, the cortisol in the circulating blood then acts on the hypothalamus to govern the output of CRF. The hypothalamus sets the desired level of cortisol in the blood. If the cortisol concentration rises above this, the outputs of CRF and ACTH and hence of cortisol are reduced. If the cortisol concentration in the blood falls below the desired level, the outputs of CRF, ACTH and therefore of cortisol are increased. The mechanism therefore tends to maintain the cortisol concentration in the blood constant.

POSITIVE FEEDBACK
With positive feedback, a disturbance leads to events which further increase the disturbance. This increased disturbance activates the positive feedback mechanism still more and so the state of the system changes very rapidly

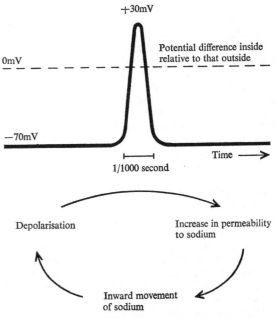

Figure 2.6 The nerve impulse and the positive feedback mechanism which produces the rapid potential change.

indeed. The positive feedback principle is perhaps more familiarly known as the vicious circle. Positive feedback mechanisms are unstable and are therefore rare in biology.

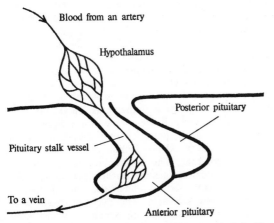

Figure 2.5 The hypothalamus, the pituitary and their blood supply.

However, there are at least three good examples. The first of these is the nerve impulse. If the potential difference between the inside and outside of a nerve fibre is measured, the inside is found to be about 70 millivolts (mV) negative to the outside. This is known as the resting membrane potential because it remains steady as long as the nerve is inactive. Before the positive feedback mechanism can be understood, two other facts must be known: these are that the resting membrane is effectively impermeable to sodium ions and that the concentration of sodium ions in the fluid outside the fibre is very much greater than the concentration in the fluid inside the fibre. There is therefore a large concentration gradient which would push sodium rapidly into the fibre if the membrane were permeable to it. If an electric shock is given to the nerve fibre, a nerve impulse may be initiated by the following sequence of events:

1. The shock reduces the potential difference between the inside and outside of the fibre. It is said to depolarise the fibre.

2. This depolarisation makes the nerve fibre membrane slightly permeable to sodium. Sodium ions therefore move into the fibre down their concentration gradient.

3. Since the sodium ions are positively charged, their inward movement further reduces the internal negativity of the fibre.

4. This further depolarisation increases the permeability to sodium still more. Sodium ions rush in even more rapidly, this produces a further depolarisation and so on.

5. Eventually sodium rushes in so rapidly that the inside of the nerve fibre actually becomes positive to the outside. The movement stops only when the concentration gradient pushing sodium in is balanced by the electrical force pushing sodium out. (Like charges repel one another, and once the inside of the fibre becomes positive it will tend to repel the positively charged sodium ions.)

However, if the membrane were to remain positively charged inside, the nerve would be inexcitable and could conduct no further impulses. There is therefore a cut off point. For reasons which are as yet poorly understood, after about one-thousandth of a second the membrane again becomes effectively impermeable to sodium. The membrane potential returns to its original level and the sodium ions are pumped out of the fibre. This demonstrates that positive feedback systems can be tolerated in biology only if at some point there is a cut out which breaks the cycle and returns the system to its former state. If this were not so, the positive feedback system could operate once only and that would not be much use to any animal.

The second example of positive feedback is much less well known and is still the subject of intensive research. In female mammals, the release of eggs from the ovary

(ovulation) seems to be brought about by a sudden surge in the output of one of the gonadotrophic hormones, luteinising hormone (LH). The output of LH both before

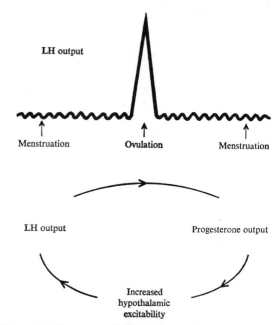

Figure 2.7 The output of LH during the female sexual cycle.

and after ovulation is low, but at the time of ovulation there is a sudden and spectacular rise to a peak of hormone output followed by an almost equally rapid fall. In fact, the general shape of a graph showing LH output during the female sexual cycle does not look unlike the shape of an action potential. Whenever anything rises as dramatically as the output of LH at ovulation, it is wise to look for a positive feedback mechanism. This is especially true when the rapid rise is followed by a rapid fall indicating the operation of a cut out. There is much argument about the detailed explanation of this phenomenon which is further discussed in chapter 10.

The third example also comes from the field of reproduction. In a pregnant woman once labour starts, part of the fetus (usually the head) presses down upon the lower part of the uterus known as the cervix. The cervix is richly supplied with sensory nerves which send impulses along a pathway to the hypothalamus. The result of activation of these nerves is the release of a hormone known as oxytocin from the posterior pituitary gland. Oxytocin circulates in the blood and stimulates the muscle of the uterus to contract. The contraction forces the fetus against the cervix causing further activation of the cervical sensory nerves which therefore initiate the release of more oxytocin. The cycle proceeds with a crescendo of increasing uterine activity until the baby is expelled from the uterus and the positive feedback cycle is automatically cut off.

In summary, if in biology one finds that a factor remains constant even when the organism is exposed to disturbing forces, one should look for a negative feedback system controlling that factor. If on the other hand the level of some factor changes extremely rapidly one should suspect the presence of positive feedback, particularly if a cut-off device is present. Even without knowing anything further, one therefore has a useful guide to research.

THE VALUE OF DISTURBANCE DETECTORS

It was earlier pointed out that if a control mechanism is to work effectively it must be supplied with information by what is known as a misalignment detector. This picks up the discrepancy between the forces tending to push the system in one direction and those tending to push it in the other. It does not measure the intensity of the forces which are tending to disturb the system. Thus the thermometer of a household thermostat is put *not* outside the house but in the room whose temperature is to be maintained at a steady level. There it measures the result of the interaction between the cold stress and the response of the heating system. Thus also carbon dioxide receptors are not put on the venous side of the circulatory system where they could measure the amount of excess carbon dioxide entering the blood from the peripheral tissues. Instead they are put on the arterial side where they monitor the result of the interaction between the entry of carbon dioxide into the blood in the tissues and its removal by the lungs. Long experience with physical control systems has demonstrated that systems which depend on misalignment detectors are stable while those which depend on disturbance detectors (detectors which measure the stress imposed on a system) are unstable and hopelessly unreliable.

However, although disturbance detectors are unsatisfactory when employed alone, there are many uses for them in both physical and biological control systems. If used in conjunction with misalignment detectors they can greatly increase the speed of response of a control mechanism. They can reduce the delay which inevitably occurs between a change in the system and the response to that change. For example, a thermostat could be made more effective if instead of a simple misalignment detector inside the house, there was in addition a disturbance detector on an outside wall. If the weather suddenly became cooler, this disturbance would be picked up by the outside thermometer before any alterations in the internal temperature could take place. If this information about the increased cold stress were then fed into the control system, the intensity of the heating could be gradually increased in anticipation of the expected fall in internal temperature. The information from the misalignment detector inside the house would then be used to make the final slight alterations in the heating system.

It is important that the disturbance detector should not dominate the control system. The final say in what happens to the intensity of heating should always depend on the misalignment detector. It is therefore clear that the control system should be so designed that it takes note only of changes in the behaviour of the disturbance detector. In the steady state, the information from the disturbance detector should be ignored. In contrast the heating should be controlled both by steady state and by changing information coming from the misalignment detector since the aim is to keep the reading of this detector constant. Furthermore, a disturbance detector should not produce too large an effector response: if it did, if the heating system were turned on full blast by a sudden fall in outside temperature, the temperature might well overswing and rise above the desired level. The link between the disturbance detector and the control system must be such that moderate effector changes in the right direction are produced, the final adjustment being dependent on information from the misalignment detector.

What would happen to such a thermostatic control system if one or other of the thermometers were destroyed? First consider the disturbance detector on the outside wall. Suppose that it is so arranged that the higher the outside temperature the more powerful is the signal it sends back to the control system. If the temperature falls the power of the signal will be reduced. If the disturbance detector suddenly stopped sending back information, the control system might therefore interpret this as a sudden fall in outside temperature and increase the output of the heating system. But within a short time the temperature inside the house would rise and the heating would be cut back to its original level as a result of the behaviour of the dominant internal misalignment detector. Loss of the outside thermometer would therefore produce no change in the temperature around which the system was being regulated and the internal temperature would still be relatively stable. In the steady state the temperature would be precisely the same as it would have been had the disturbance detector been functioning. The only difference will be in the nature of the response to changes in outside temperature. When the surroundings became colder, the heating system response will begin only when the change has produced a fall in internal temperature. The internal temperature will therefore fall further before it can be counteracted by an increased intensity of heating. Similarly if the environment becomes hotter the internal temperature will rise higher before the heating is turned down. Oscillations around the set temperature will therefore be wider than before.

The results will be quite different if the internal thermometer fails. The control system will interpret the sudden loss of information as a very rapid fall in internal temperature. The heating system will be turned up to its maximum level and since the control system is sensitive to the steady state discharge of the internal thermometer, this situation will continue even though the temperature inside the house rises rapidly. The loss of the misalignment detector will render the thermostat system useless.

HUMAN TEMPERATURE REGULATION

The idea that both disturbance and misalignment detectors can be useful in control mechanisms is a very fruitful one in biology. Consider first the body's own temperature regulating mechanism as it operates in warm-blooded creatures. Which part of the body is most nearly maintained at a constant temperature? It is certainly not the skin for the skin temperature may fluctuate widely depending on the outside conditions. In fact the arterial blood has the most nearly constant temperature of all the body tissues and it therefore seems reasonable to suggest that it is the temperature of the arterial blood which the regulating systems attempt to keep constant. The relatively constant temperature of many other parts of the body merely indicates that they are well-supplied with arterial blood.

Misalignment detectors should therefore be located where they can monitor the temperature of arterial blood. Since the arterial blood supply to the nervous system is of over-riding importance, it might be expected that these detectors would be situated in the brain. This expectation has been found to be true and there is abundant evidence that the arterial temperature receptors are situated in the hypothalamus. Disturbance of hypothalamic function in humans because of vascular disease or by a tumour can produce severe derangement of temperature regulation. The body temperature may remain persistently high or persistently low.

But as common experience indicates, temperature receptors are not only found in the brain. The skin is very obviously temperature sensitive and rapidly detects changes in the temperature of the environment. However, the skin receptors adapt quickly to new situations and in the steady state one is unaware that they are functioning. But the receptors do warn of temperature changes and they are ideal disturbance detectors. Many experiments have shown that they can function in this way. For example, if an animal is exposed to radiant heat the blood flow to the skin is increased and sweating begins before any change can be detected in the arterial blood temperature. The information from the skin receptors enables the control system to anticipate a change in the blood temperature. Human beings and animals exposed to a hot sun will seek shade and thus reduce the heat stress long before the blood temperature has altered. In both cases the skin receptors act as disturbance detectors warning the control system that unless the rate of heat gain is reduced and the rate of heat loss is increased, the arterial blood temperature will rise. If the blood temperature rises because the measures set in motion to lower body temperature have been too energetic, this will be detected by the central hypothalamic receptors and the final fine control will operate. The control of arterial blood temperature would be less precise and the oscillations would be greater without the existence of disturbance detectors on the body surface.

What will happen if the two sets of receptors, central brain and superficial skin, provide conflicting information? If the central temperature receptors are behaving as true misalignment detectors while the skin ones are simply disturbance detectors, the information provided by the central ones should be dominant. Exper-

iments of this type have been done. The arterial blood can be cooled either by placing a large volume of iced water in the stomach or by actually infusing cold salt solution into a vein. At the same time the skin can be kept very warm by keeping the subject of the experiment in a hot room. Under these conditions, even though the skin is warm, sweating falls to very low levels, skin blood flow is reduced and even shivering may occur. If the reverse experiment is done and the arterial blood is warmed while the skin is kept cold, sweating and skin blood flow both increase. Thus what actually happens is precisely what one would expect if the skin receptors were acting merely as disturbance detectors while the arterial blood receptors were acting as dominant misalignment detectors.

DISTURBANCE DETECTORS IN RESPIRATION

Another good example of the use of disturbance detectors can be seen in the control of the breathing during exercise. As mentioned earlier, once exercise is being done at a steady rate, the rate of ventilation seems to be determined by the rate of carbon dioxide production. As a result of the operation of arterial misalignment detectors for carbon dioxide, the rate of ventilation is so increased that the increased rate of carbon dioxide removal is just balanced by the increased rate of production. But when a person at rest suddenly begins to take exercise, it is some time before the increased muscular activity causes an increased outpouring of carbon dioxide into the venous blood. If the misalignment detectors for carbon dioxide were the only sources of information for the control of ventilation, there would be a finite time lag between the beginning of the exercise and the rise in the ventilation rate. What actually happens? Suppose we study the ventilation rate in someone sitting on a bicycle so mounted that it remains stationary when the pedals are turned. Suppose also that it is possible to apply a brake to the wheels so changing the resistance

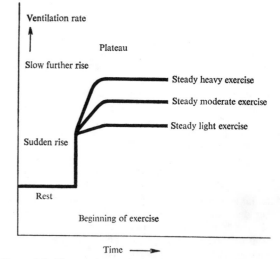

Figure 2.8 Changes in ventilation rate on beginning exercise and on reaching the steady state.

against which the pedals must be pushed. When a person who has been at rest suddenly begins pedalling, the ventilation rate rises sharply long before any change can be detected in the rate of carbon dioxide entry into the blood. This sudden increase is then followed by a further, slower rise until a steady rate of ventilation is reached which copes satisfactorily with a steady rate of exercise. If a series of separate tests is done in which the subject is asked to pedal against different resistances, it is found that the sudden initial increase is roughly the same whatever the resistance. However, the final plateau depends on the intensity of the work being done, the harder the work, the greater is the second slower rise.

Experiments in animals have revealed that, as might be expected, the sharp initial rise does not depend on chemical changes in the blood. Instead it depends on the activation of mechanical receptors in the exercising limbs, particularly those in the joints which signal to cularly those in the joints which signal to the brain when and how rapidly the joints are moving. Therefore when a limb is moved, whether actively or passively or against a light or heavy load, the joint receptors warn the control centres in the brain that exercise is taking place and that the output of carbon dioxide from the muscles is likely to increase. The joint receptors are acting as typical disturbance detectors. In anticipation of the increase in carbon dioxide production, the control system produces an increase in ventilation rate. If the exercise is very light and the increase in carbon dioxide output is small, the initial increase in ventilation approximately compensates for it and there is little further rise in ventilation. On the other hand, if the exercise is heavy, the initial increase in ventilation will not cope with the large amounts of carbon dioxide produced. The arterial carbon dioxide level will tend to rise and this will stimulate the receptors

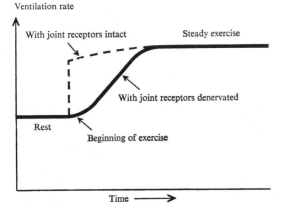

Figure 2.9 The effect of denervation of the joint receptors on the changes in ventilation rate which occur at the beginning of exercise.

which will send the information to the control centre in the brain. The brain will then order an increase in ventilation rate which will continue until the increased rate of removal of carbon dioxide just balances the increased rate of production and the arterial level returns to normal.

If the joint receptors are destroyed or anaesthetised, one would not expect the final response of ventilation to steady state exercise to be altered. One would expect the initial response to change to be different. This is in fact what occurs. Without the mechanical receptors which inform the brain that limbs are moving there is no initial sudden rise in ventilation rate. Instead, as the rate of carbon dioxide output increases, there is a steady but slower rise in ventilation rate. The final plateau of ventilation rate is the same as would have occurred had the joint receptors been functioning normally. Therefore as would be expected from their function as disturbance detectors, the limb receptors influence the rate of response to change but not the magnitude of the final response.

APPETITE AND THIRST

The mechanisms controlling appetite and thirst provide further biological examples of situations where both disturbance and misalignment detectors are important. The sensation of thirst depends primarily on minute increases in the ionic concentration of the blood which indicate that the body is being depleted of water. The detection of these changes depends again on receptors in the hypothalamus which monitor the composition of the blood flowing through the brain. If the ionic concentration rises above normal because of loss of water then a sensation of intense thirst results. The situation can be produced artificially by injecting minute amounts of concentrated salt solution into the carotid arteries which supply the brain.

Suppose that the sensation of thirst is investigated in a dog which has previously been subjected to an operation in order to make a gap (fistula) in its oesophagus. Any food or drink that is taken in through the mouth therefore escapes from the gap in the oesophagus without entering the stomach. If the dog is deprived of water for a couple of days it will become very thirsty and blood sampling will reveal that the ionic concentration of the blood is above normal. What then happens if water is provided? The dog will drink voraciously and will continue drinking until it is exhausted because the water does not reach the stomach and intestines and therefore does not enter the body. On the other hand, suppose that in a thirsty dog a large volume of water is poured into the stomach before the dog is allowed to drink. When the animal is allowed access to water it will take just a few mouthfuls and then stop. If a blood sample is taken at this point it reveals that its sensation of thirst has vanished long before all the water in the stomach has been absorbed into the body and long before the chemical changes in the blood have been reversed to normal. Sensory receptors in the wall of the stomach which are stretched when the stomach is filled act as disturbance detectors informing the central control centre that the water deficit is likely to be made good. When after 30–60 min all the water has been absorbed, the more precise blood sampling receptors in the hypothalamus which act as misalignment detectors assess whether or not the deficit has been accurately replaced. If it has not the sensation of thirst may return.

The disturbance detectors are very important in the regulation of the body water content. Suppose that the misalignment detectors in the hypothalamus were the only receptors supplying information to the control system. Suppose that in order to abolish the sensation of thirst, drinking had to continue until the chemical changes in the blood were fully reversed to normal. After all ordinary degrees of dehydration, the amount of water required to replace the loss can be swallowed in a few minutes. But it may be thirty minutes or more before that water is absorbed from the gut and the chemical changes are corrected. If a human being or animal remained thirsty all that time, far more water would be swallowed than was actually required. By the time the chemical composition of the blood had returned to normal, large amounts of water would be left in the gut. The absorption of this water would dilute the blood and the kidney would have to excrete the excess in the form of urine.

Much the same considerations apply to appetite. The sensation of hunger almost certainly depends on subtle alterations in the carbohydrate and fat content of the blood which are by no means clearly understood as yet. Again the receptors which detect these changes seem to be in the hypothalamus. It is a matter of common experience that no matter how hungry one may be the sensation of hunger is greatly diminished by just a few mouthfuls. The sensation vanishes while most of the food is still in the gut being digested and long before any biochemical changes could have been reversed. Animal experiments have demonstrated that the sensation of hunger can be considerably reduced simply by stretching the stomach, even if the stretching is done by material which has no food value. Again the disturbance detectors are of obvious significance. If we ate until the biochemical changes in the blood were completely reversed, all meals would be orgies lasting several hours. When we had reached the stage of being no longer hungry, the gut would still be loaded with food and there would be no need to eat again for many hours more. We might therefore eat once per day or even once every several days. It is possible that appetite control in animals such as snakes may work in this way.

THE INFLUENCE OF THE TYPE OF STRESS ON THE DESIGN OF A CONTROL SYSTEM

With some control mechanisms, outside stresses which tend to disturb the system always tend to push the system in one direction. With other control mechanisms, there may be several different types of outside stress, some tending to push the system in one direction and some tending to push it in the other. Control systems to cope with the second situation must be much more complex than those which can cope with the first. Concrete examples may help to demonstrate this.

Suppose that one built a submarine chamber beneath the Arctic ice cap. The chamber would be surrounded by water whose temperature remains at about 4 °C all the year round. Suppose that one wanted to maintain the temperature inside the chamber at 20 °C. The only source of heat supplied to the chamber would be the heating system actually installed in it. If as a result of the operation of the heating system the chamber became slightly too hot, the effect of the water outside would automatically cool it down as soon as the heating was turned down. A very simple control system could therefore maintain the temperature of the chamber constant all the year round.

In contrast, the problems of controlling the internal temperature of a house placed in the middle of a continent in one of the temperate regions of the world, say in Central Europe or the American Mid-West, are quite different. In winter the difficulties are similar to those experienced in the Arctic. All one needs is a heating system. If the heating system makes the house too hot, then all that is required is to turn the heating down. The cold outside will then lower the internal temperature in a satisfactory way. But this simple mechanism will be quite useless in the summer when the temperature outside may be much hotter than the temperature which is required inside. Simply turning off the heating will not then bring down the internal temperature to the required value. An active refrigeration system for cooling the air inside the house is required. Thus if the stress is always in one direction, one type of effector mechanism will suffice to achieve satisfactory control. If outside stresses can push the system either way from the desired set level, then at least two different effectors are required.

Some problems of physiological control are similar to those of heating a chamber in the Arctic sea: the stress is always in one direction. For example, consider the control of the secretion of ACTH from the anterior pituitary. This depends only on the amount of CRF reaching the pituitary from the hypothalamus. ACTH output cannot be stimulated in any other way. If the pituitary is removed from its position at the base of the brain and transplanted to a quite different part of the body, it ceases to be exposed to the action of releasing factor in the normal way. The output of ACTH falls to zero and there are no external stresses which can increase it. Thus when the pituitary is in its normal position, by varying the output of releasing factor the output of ACTH can be varied over its full range. The one mechanism for controlling ACTH output will suffice and there is no need to invoke any other. Similar considerations apply to the control of the output of thyroid hormone by TSH. If TSH output is zero then the output of thyroid hormones is also zero. Again by varying the TSH output, the output of thyroid hormone can be varied over the full range required.

In contrast, there are several biological control mechanisms where one type of effector control is not sufficient. The problem of regulating blood glucose level is one such case. After a meal, when the food is being absorbed into the body from the gut, the blood glucose concentration tends to rise to very high levels. If the concentration is to be maintained within normal limits, it is essential to have a mechanism for bringing the level down. This is done very effectively by insulin. On the other hand during exercise the muscles consume glucose extremely rapidly and there is a danger that the blood glucose level may fall. Since the brain depends almost entirely on glucose for its energy supply, low glucose levels in the blood could clearly produce a dangerous situation with a risk of loss of consciousness. It is therefore not surprising that in order to counter this danger there are at least four mechanisms, all acting in different ways, which can raise the blood glucose level and maintain the supply to the brain. Growth hormone from the pituitary, cortisol from the adrenal cortex, glucagon from the pancreas and adrenaline from the adrenal medulla can all act to raise blood glucose. Thus because blood glucose may either tend to rise abnormally high or to fall abnormally low in different situations, there are active mechanisms for raising it and quite different ones for lowering it.

The temperature regulation of the body provides another instance. If the body temperature falls in a cold environment, heat production can be increased by exercise and by shivering and heat loss can be cut down by reducing the blood flow through the skin. If the environmental temperature is high and the body temperature rises, then heat production can be cut down by physical rest and heat loss can be increased by active sweating and by an increase in the skin blood flow. The body must be able to resist disturbance of temperature in both directions and in each case different effector mechanisms are involved. The body possesses both active heating and active refrigeration systems.

SYSTEMS WITH MULTIPLE DETECTORS AND MULTIPLE EFFECTORS

Biologists, working with organisms which contain by far the most complex computers known, often seem to take an unduly simple-minded view of the systems which they are studying. Their model of a control mechanism is all too often the household thermostat in its most basic form. This consists of one temperature measuring device, one control device, sensitive to changes in the thermometer reading, and one heating system which responds to the control device in an appropriate way.

Suppose you were in charge of the design of a temperature controlling system for a space capsule which was to be landed on Mars and which was going to be subjected to extremes of heat and cold. It would no doubt be very, very clear to you that the lives of the men inside the capsule would depend on the effective functioning of your system. If your system failed they would either freeze or fry. Would you be content therefore with installing in the capsule one thermometer, one controlling device and one heating and cooling system? I suggest that you would not. Your measuring system would consist of at least two thermometers and possibly more so that if one failed for any reason, the others could quickly take over and temperature regulation would continue to be effective. You might also install at least two controlling devices and two effector systems so that if one failed another could take over.

I venture to suggest that many of the body's controlling mechanisms, shaped by millions of years of fighting for survival in evolution are no less sophisticated. Most of the important factors in the body which must be precisely controlled are probably monitored by sense organs in several different sites. In most cases, too, there are probably several different mechanisms whereby a desired effector response can be brought about. The control of arterial blood pressure is an excellent illustration of what I mean. It is essential that under normal conditions the pressure of the blood in the arteries should be regulated within fairly narrow limits. If the pressure is too low, there may not be enough force to supply blood to the various organs: as a result the organs may die or be permanently damaged by the lack of blood. If the blood pressure is too high blood vessels in some organs may burst open and a great strain may be put on the heart. Because the control of blood pressure is so important one might expect that there would be several different receptors supplying information to the brain about it and several different effector mechanisms for controlling it.

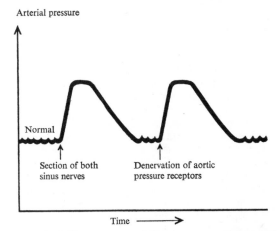

Arterial pressure

Normal

Section of both sinus nerves

Denervation of aortic pressure receptors

Time

Figure 2.10 The sequence of blood pressure changes after destruction first of the sinus nerves and then of the aortic nerves.

This is indeed the case. There are pressure receptors in the walls of many different blood vessels. When the pressure inside the vessel rises, the walls are distended and the receptors are stretched. The higher the pressure the more impulses are fired by the receptors and transmitted to the brain. The latter therefore receives a barrage of information about the pressure of the blood in many different arteries. The most important of these pressure receptors seem to be the ones in the carotid artery in the neck in the region of the artery known as the carotid sinus. They are in a good position to measure the pressure of the blood in the arteries which supply the brain. They send the information about pressure to the central control mechanism along what is known as the sinus nerve. What therefore happens if the sinus nerves on both sides of the neck are cut, thus stopping information about pressure reaching the brain via this pathway? Under normal circumstances, a fall in the number of nerve impulses travelling along the sinus nerve would indicate a fall in blood pressure. Initially in the experiment the brain is fooled. It does interpret the cutting of the nerve as a fall in blood pressure. In a vain effort to restore the discharge in the sinus nerves to normal levels, the brain sets in motion effector changes which raise the arterial pressure well above normal. The pressure remains high for a time but then gradually returns to normal levels even though the important sinus nerves have been destroyed. This return to normal arterial pressure is not difficult to understand.

There are many other arterial pressure receptors supplying information to the brain. These continually signal that the arterial pressure is above normal. The control mechanism will rapidly be made aware of the discrepancy between the information from the sinus nerves and that from other pressure receptors.

Furthermore, assuming that the brain is a sophisticated computer, it will no doubt soon realise that there is no correlation between its attempts to raise arterial pressure by altering effector mechanisms and the activity in the sinus nerves. When the brain sets in motion mechanisms which raise arterial pressure, the sensory discharge in the pressure receptors normally rises accordingly. In this experiment nothing happens in the sinus nerves when the pressure is raised but the discharge from other receptors increases as usual.

The brain therefore ignores completely the lack of activity in the sinus nerves and relies entirely on information from elsewhere. The cycle is repeated if what appear to be the second most important pressure receptors, those in the aorta (the main artery which comes from the heart) are then denervated. Again the pressure rises initially but again it falls after a period as the normally less important pressure receptors in other arteries take over. It is important to note that if the receptors of secondary importance are destroyed while the sinus receptors are left intact, the blood pressure does not usually change. But that does not mean that the secondary receptors are of no significance. Like the secondary thermometers in the thermostat system of a spacecraft, their importance is revealed only when the primary receptors fail.

In controlling arterial pressure, the body is equally versatile on the effector side. The maintenance of a normal arterial pressure depends on the fact that the heart pumps out blood into a system of tubes which offers resistance to the flow of that blood. If the resistance remains constant, the more blood that is pumped out per minute the higher will be the pressure. The pressure therefore depends partly on the cardiac output, the amount of blood pumped out by the heart every minute. On the other hand, if the cardiac output remains constant but the resistance offered by the vessels becomes greater, the pressure will again rise. The reverse changes will produce a fall in arterial pressure. The resistance offered by the system of tubes can be altered because the walls of the small blood vessels are muscular. If the muscle in a vessel wall contracts, the diameter of that vessel becomes less

and the resistance to blood flow increases. If the muscle relaxes, the resistance to flow will fall: if the cardiac output then remains constant, the arterial pressure will also fall.

And so the arterial pressure can be altered either by changing the resistance or by changing the cardiac output. But the mechanism is considerably more versatile even than that. The cardiac output itself can be altered in many different ways. To name but three of these, the output can be increased by an increase in the activity of the excitatory sympathetic nerves to the heart, by a decrease in the activity of the inhibitory vagal nerve fibres or by an increase in the concentration of the hormone adrenaline (epinephrine) in the blood. Similarly the resistance offered by the blood vessels can also be altered in many different ways. Therefore if only one of the effector control mechanisms is damaged, any effect on the level of arterial pressure is likely to be only slight and temporary. The brain has at its disposal so many methods of altering both cardiac output and resistance that only if the majority of them cease to function, is blood pressure likely to be permanently effected.

Thus in many biological situations only the naive would expect any permanent alteration if only one of several possible sources of information or only one of several possible effector mechanisms is destroyed. Despite this, biological history contains many examples of false reasoning based on the argument that a sensory receptor is insignificant in effect if its destruction is followed by virtually normal functioning of the body. That is rather like saying that if the temperature of a spacecraft remains steady after one of two available thermometers in the control system has been destroyed, then the destroyed thermometer was functionless and unimportant. This is clearly nonsense because although the system still functions normally, one line of defence has gone and the spacecraft is less well placed to cope with further damage. The possession of many receptor and many effector mechanisms dealing with a single factor is a simple precaution of defence in depth forced upon successful organisms by many years of evolution.

With arterial pressure, although there are many receptors, theoretically the regulating system could function only with one since the pressure of the blood is essentially the same throughout the arterial tree. A similar thing is true of mechanisms which control the chemical composition of the blood. The concentrations in the blood of most substances are effectively identical throughout the vascular system. Only one receptor could do the job of monitoring the blood although in practice there may be several. However, some things which are kept constant by their very nature require the collection of information from many different sources. For example, in any one individual the volume of blood is kept remarkably constant. Yet there can obviously be no single simple way of measuring the volume of blood stored in such an extremely complicated system of tubes as the vascular system. The accurate system of blood volume control must therefore depend on collecting information about the volume of blood in arteries, veins and in small vessels in many organs. Only by collecting information from many, many receptors and integrating that information in the brain can a reliable picture of the blood volume be compiled. If this is so, it is unlikely that an experiment which investigates the effect on the control of blood volume of the destruction of just one or two likely receptors will have any positive result. All the other receptors involved will still be signalling 'situation normal' and the brain will conclude that even if the cessation of information from the destroyed receptors means anything at all it cannot be very significant. Therefore if one denervates an organ which may be playing a role in the collection of information about blood volume and nothing happens, then one is not entitled to conclude that that receptor is not involved in volume control.

ALTERATION OF THE BEHAVIOUR OF CONTROL MECHANISMS
With most thermostats, even ones of very simple design, it is possible to alter the setting around which the control mechanism operates. The householder can decide whether he wishes the temperature of a room to be maintained at 10 °C, 15 °C or 20 °C by turning a dial which alters the point at which the heating and cooling systems turn on and off. Many of the body's control mechanisms appear to have similar devices which may either alter the behaviour of a sensory receptor or, without altering the behaviour of the receptor, may alter the way in which the central control mechanism responds to the information provided.

DEFAECATION
A mechanism which clearly illustrates the two ways in which a control system can be altered is that of the emptying of the large bowel (defaecation). The wall of the large bowel contains sensory receptors which are sensitive to stretch. The receptors are arranged so that they are in series with muscle fibres in the bowel wall. There are therefore two ways in which the sense organs can be stretched.

1. By an increase in the volume of the contents of the bowel.

2. By contraction of the muscles in the wall of the bowel.

When the receptors are stretched, the information is sent back to a control centre in the spinal cord. When the receptor discharge reaches a critical level the control centre orders emptying of the bowel. In infants and in paraplegics (patients who have had their spinal cord cut across), the mechanism operates in this simple way. As the large bowel becomes loaded with faeces the receptors are stretched more and more. At a certain level of discharge the effector mechanism is triggered and defaecation follows automatically. Clearly if this simple mechanism operated in normal adults the consequences could be very unpleasant and, as everyone knows,

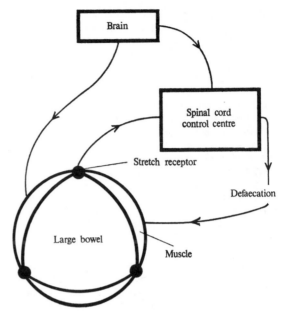

Figure 2.11 The mechanism of voluntary control of defaecation.

defaecation is largely under voluntary control. This control is mediated by nerve fibres which descend the spinal cord from higher regions of the brain and which produce two different effects.

1. They govern the state of contraction of the muscle fibres linked to the sensory receptors. If the muscle fibres contract the sensory discharge will increase, while if they relax the sensory discharge will fall even if the volume of the gut contents does not change.

2. They govern the sensitivity of the control centre in the spinal cord. If the sensitivity is reduced, even a high level of receptor discharge will not cause defaecation. If the sensitivity is increased, a low level of sensory activity will bring about emptying of the bowel.

Thus in practice the timing of voluntary defaecation is controlled in two ways. When defaecation is desired, the muscles in the bowel wall contract and this increases the sensory discharge. Simultaneously the sensitivity of the control centre is increased, so making it easier for the sensory discharge to fire the defaecation response.

FEVER
The condition known as fever provides another good instance of an alteration in the behaviour of a control system. Fever seems to occur whenever cells are being destroyed whether by infection, injury, cancerous growth or in any other way. Such cell destruction stimulates some of the white cells of the blood (the neutrophil leucocytes) to release a substance known as pyrogen. Pyrogen acts on the hypothalamus to alter the setting around

which the hypothalamic thermostat operates. The function of the response is not clear and as yet there is little evidence that it helps to protect the body against disease. Nevertheless it does occur and its mode of operation is well known.

The arterial blood temperature is normally regulated around the level of 37 °C. If it rises above this, the skin blood flow increases and sweating occurs in order to bring it back to normal. If the temperature falls below 37 °C, heat production may be increased by shivering and heat loss reduced by cutting down skin blood flow until the temperature returns again to normal.

What then happens at the beginning of fever? Even though the temperature is normal, heat production is increased by shivering and heat loss is reduced by cutting down the skin blood flow. The body temperature therefore rises until it reaches a new equilibrium level, say 39 °C. If then the body temperature is artificially reduced to 38 °C by cold sponging or by the infusion of cold saline, shivering sets in and the temperature soon rises again. Yet in a normal individual, a temperature of 38 °C would produce quite the opposite reactions, namely sweating and increased skin blood flow. The pyrogen has clearly shifted the setting of the thermostat. When the pyrogen level falls at the end of fever, the setting reverts to 37 °C and the temperature is brought back to normal by excess sweating and increased skin blood flow.

CORTISOL

Yet another example may be taken from the field of endocrine physiology. ACTH is released from the anterior pituitary under the influence of its releasing factor, CRF. The amount of CRF produced by the hypothalamus depends on the amount of cortisol in the circulating

blood. If the cortisol concentration rises above a certain critical level, X, the amount of releasing factor put out falls. If the cortisol level falls below X, the CRF output rises. The changes in CRF, acting via ACTH, ultimately produce corresponding changes in the rate of the synthesis of cortisol by the adrenal cortex. Therefore the level of cortisol in the blood remains approximately constant. But the setting of the cortisol–CRF link can be altered by several other factors. One of these is any form of stress, whether physical or psychological, imposed on the individual. Information about the stress is conveyed to the hypothalamus where it leads to a reduction in the sensitivity of the CRF response to cortisol. Even though the concentration of cortisol in the blood may be X, this is now insufficient to control the output of CRF. The rate of CRF release is therefore increased, ACTH and hence cortisol outputs rise, and eventually, at a new level of blood cortisol $(X+Y)$, the output of CRF becomes stabilised. In the new situation if the cortisol level rises above $(X+Y)$ the output of CRF decreases: if the level falls below $(X+Y)$ the output of CRF increases. At the end of the period of stress, the sensitivity of the hypothalamic mechanism returns to its old level and the concentration of blood cortisol falls back to X.

CONCLUSION

This chapter has given an outline of the principles of operation of most of the important types of control system which operate in the mammalian body. The remaining chapters of the book give more detail about each of the systems and during the course of them you will come across many more examples of the importance of control mechanisms in the maintenance of life.

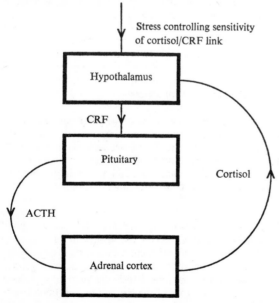

Figure 2.12 The modification of the feedback link between cortisol and CRF by stress.

3 Biochemistry

Biochemistry is the study of the chemistry of living organisms, of the ways in which food is used to serve all the many needs of the body. Biochemistry is closely connected with nutrition, the study of the types and amounts of various materials required in the diet. Biochemistry is also inextricably intertwined with endocrinology, the study of hormones, for most of the hormones exert their actions by altering the behaviour of chemical reactions within the body.

The central problem in biochemistry is that of the supply of energy. Energy is needed for a multitude of purposes of which muscular activity is the best known. Energy is required for digestion, and for the functioning of the kidney, the liver, the brain and all the other organs in the body. Energy is also essential for the building up of the complex organic molecules of which the body is constructed.

Ultimately, most of the energy utilised on earth comes from the sun. Plants are able to tap this energy source directly by the process of photosynthesis. By using pigments, notably the green chlorophyll, plants can trap the energy of sunlight and use it to build up complex substances such as fat, carbohydrate, protein and nucleic acids. The only raw materials required are carbon dioxide, water and simple inorganic substances such as nitrates which can be extracted from the soil. In contrast to plants, animals lack pigments like chlorophyll and so they cannot directly trap the energy of sunlight and turn it to their own use. Instead they must obtain their energy indirectly by eating plants or other animals which have themselves eaten plants. The complex fats, proteins and carbohydrates which plant and animal tissues contain can then be broken down and made to yield up their energy.

THE IMPORTANCE OF ATP

Energy can be released by the breakdown of many different chemical compounds. This energy can then be used for the manufacture of other compounds, for muscular activity and for many other different purposes.

How is the link between the release of energy and the employment of that energy established?

It may be helpful to consider the supply of energy in man-made systems. Undoubtedly electricity is by far the most important type of energy used by man. Electricity can be manufactured by a number of different processes. The energy of falling water, of coal, of wood, of oil, of natural gas and of the atom can all be converted by appropriate power generating stations into electricity. In turn, this electricity can be used by both domestic and industrial consumers in a myriad different ways, for lighting, for heating, for refrigeration, for cooking, and for driving the electric motors of thousands of different types of machine, ranging from hair dryers to railway engines. Electricity is thus a most useful type of energy: it can be manufactured in many different ways and once made it can be used in many different ways.

Is there anything comparable to electricity in the body? There is. It is the substance called adenosine triphosphate or more familiarly as ATP. Like electricity it can be made in many different ways, notably by the breakdown of the carbohydrates, fats and proteins found in the food. Like electricity, the energy of ATP can be used in many different ways. It is employed in muscular contraction, in nervous activity and in the manufacture from simple substances of all the complex materials which the body requires. ATP is an almost universal substance. It is not confined to humans or even to mammals. It is the energy source which drives Olympic athletes, which lights up the tail of the firefly and which provides the electric shock of the electric eel. It is the central point in the chemistry of the body.

SOME IMPORTANT CONCEPTS AND DEFINITIONS

Metabolism is the term given to the sum of all the chemical reactions occurring at any time in the body. Anabolism indicates those reactions in which larger molecules are being built up from smaller ones: anabolism is a process which uses up ATP. Catabolism indicates those reactions in which large molecules are being broken down into smaller units: a major function of catabolism is the manufacture of ATP.

Most of the ATP in the body is obtained by the breakdown of fats, carbohydrates and proteins in the presence of oxygen. This process is known as oxidation and its main end-products are carbon dioxide and water. It is similar in principle to the process of burning which also involves the breakdown of complex materials in the presence of oxygen. But whereas burning is usually uncontrolled and the energy released by it is wasted, the process of oxidation of food in the body is carefully regulated and much of the energy released is trapped in the form of ATP. The breakdown of food materials in the presence of oxygen is known as aerobic catabolism. Very much smaller amounts of ATP are formed by the breakdown of food in the absence of oxygen. This process is known as anaerobic (without air) catabolism.

It is useful to have some measure of the total amount of energy being used by the body. The most convenient way of expressing this is in the form of calories. When food is burned it gives out the energy stored in it in the form of heat. The amount of heat released is a measure of the amount of energy made available to the body when the same amount of food is oxidised in the process of catabolism. The heat output of the food is measured as calories. One calorie is the amount of heat required to raise 1 cm^3 of water through 1 °C. The calorie itself is a very small unit and so the kilocalorie, which is 1000 calories, is more often used. In older books, calorie, with a small c means calorie while Calorie with a big C means kilocalorie. This has led to so much confusion that it seems wiser to use the scientific words calorie and kilocalorie. Therefore, when you read that such a quantity of such a food has a calorific value of 100 kcal it means that that amount of that food when burnt in the presence of oxygen will release 100 kcal of heat. When oxidised in the body it will provide an amount of energy equivalent to 100 kcal of heat.

IMPORTANT SUBSTANCES IN BIOCHEMISTRY
This section outlines the role of various important substances in biochemistry. It is intended to give a bird's eye view of the subject so that later you can see where each piece of information fits into the overall pattern.

WATER
This is the single most important substance in the body. In a lean adult male, 60% of the body weight is made up of water. In females and fat males the proportion of water is rather less. This is because fatty tissue contains very little water and so adding fat to the body adds very little water: the percentage of the body weight made up of water therefore falls. Water is important because most of the substances found within cells are dissolved in it. The blood which carries materials all around the body consists largely of water. The urine contains waste materials dissolved in water. These are just a few important examples of how important water is. It is by far the most important substance in the diet. When deprived of any other food material, any healthy individual can survive for weeks or even months. But if totally deprived of water, the healthiest of men will be dead within a few days.

OXYGEN
When substances burn in a fire, they do so because they combine with oxygen. This combination leads to the energy stored in the substances being liberated as heat. The body too uses oxygen to 'burn' food materials but it does so in a much more controlled way than in a fire. Only very small amounts of energy are released in the body by catabolism in the absence of oxygen (anaerobic metabolism).

When oxygen is used (aerobic metabolism), the ultimate end products are water and carbon dioxide, in addition, of course, to ATP. The water may amount to 200–300 ml/day and should not be forgotten when calculating how much water a patient has received or lost.

Oxygen is supplied to the body via the lungs where the air comes into close contact with the blood. Carbon dioxide leaves via the lungs, moving in the reverse direction to oxygen.

CARBOHYDRATES
Carbohydrates are substances which contain carbon, hydrogen and oxygen in the ratio 1:2:1. For example, the formula of glucose, one of the best known carbohydrates, is $C_6H_{12}O_6$, i.e. one molecule of glucose contains six carbon atoms, twelve hydrogen atoms and six oxygen atoms. The basic building blocks of which carbohydrates are made are known as monosaccharides. Monosaccharides with five carbon atoms are known as pentoses and those with six carbon atoms (like glucose) are known as hexoses. The hexoses are particularly

Figure 3.1 The chemical formulae of glucose, a typical hexose and of ribose, a typical pentose.

important in metabolism: they include glucose, galactose and fructose. Pentoses are also important since two of them, ribose and deoxyribose, are essential components of the nucleic acids.

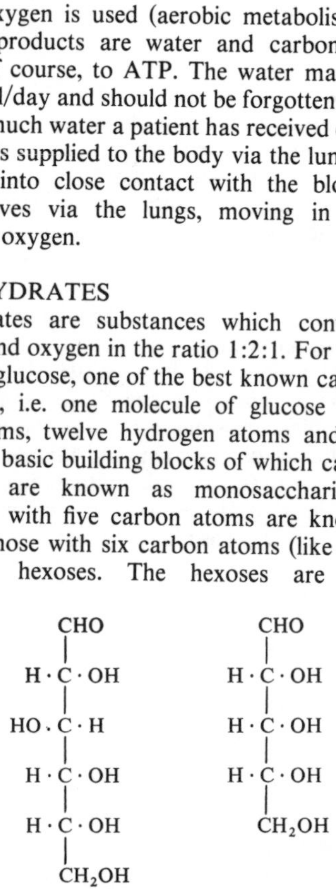

Disaccharides are also both familiar and important. As their name suggests they contain two monosaccharide molecules linked together. Sucrose (cane sugar) contains a glucose molecule and a fructose molecule. Lactose, the main carbohydrate in milk, consists of a glucose and a galactose molecule. Maltose, frequently formed in the gut by the breakdown of more complex carbohydrates, consists of two linked glucose molecules.

Polysaccharides are made up of long chains of monosaccharide molecules. Only three are of much importance. Cellulose is the tough substance of which plant cell walls are made. It cannot be satisfactorily digested by man or by meat-eating mammals. It can be used only by animals which have special digestive structures such as the rumen of the cow. Such plant-eating animals have many bacteria in their guts and it is the bacteria which initially break down the cellulose. Only when the bacteria have done their work can the animal absorb the breakdown products. Starch is another polysaccharide which is a plant product. It is the form in which carbohydrate is stored inside plant cells. Both starch and celluose consist entirely of chains of glucose molecules. Finally, glycogen is the animal equivalent of starch. It is stored in cells in liver and muscle. It too consists solely of glucose. The differences between cellulose, starch and glycogen are due to differences in the ways in which the glucose molecules are strung together.

FATS

Fats too contain only carbon, hydrogen and oxygen. However, they contain less oxygen and hydrogen for each carbon molecule than do the carbohydrates. The fats are chemically a mixed group of substances all of which share the property of being soluble in such substances as chloroform and ether. Some fats can dissolve in water as well but most tend to be water insoluble. Perhaps the most important groups of fats in the body are:

1. *Glycerides*

Glycerides are built up by the combination of fatty acids with glycerol (popularly known as glycerine). Glycerol

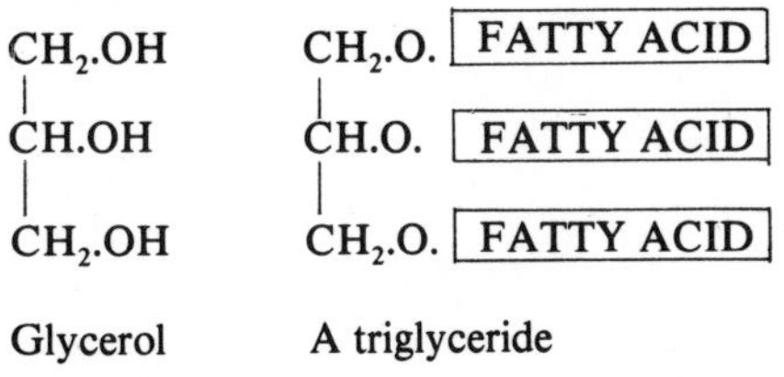

Figure 3.2 The formulae of glycerol and a triglyceride.

can combine with fatty acids by means of its three hydroxyl (OH) groups. When one of the hydroxyl groups is combined with a fatty acid, the resulting compound is said to be a monoglyceride. When two hydroxyl groups are combined with fatty acids the result is a diglyceride and when three are combined the result is a triglyceride.

The fatty acids themselves (sometimes known as free fatty acids, FFA, or non-esterified fatty acids, NEFA) consist of chains of carbon molecules with hydrogen ions attached to them and a carboxylic acid group (COOH) at one end. They come in two main forms known as saturated and unsaturated. One theory which attempts to account for the present high incidence of heart disease in developed societies suggests that it is caused by a change in the balance of saturated and unsaturated fatty acids in the diet. With saturated fatty acids, each carbon atom has two hydrogen atoms attached to it and so all the four valency bonds of carbon are used up, hence the word 'saturated'. With unsaturated acids, some carbons have only

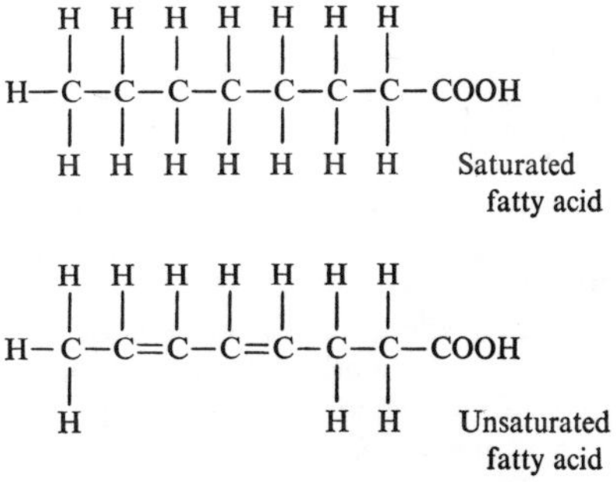

Figure 3.3 Saturated and unsaturated fatty acids.

one hydrogen atom attached to them. There is thus a spare valency bond which is usually attached to the next carbon atom giving a double bond. Unsaturated fatty acids tend to be found in liquid fats such as palm, corn and soya-bean oils. Saturated acids predominate in solid fats. Margarine, for example, is made from oils. During its manufacture these oils are made to react with more hydrogen. This makes them more saturated and hence more solid.

The main function of the glycerides in the body is to act as a supply of energy. Some of the unsaturated fatty acids, known as essential fatty acids, have a vitamin-like function.

2. *Steroids*

These are substances which have four rings of carbon atoms. Cholesterol is perhaps the best known steroid as it is found in the blood and high blood levels appear to be associated with heart disease. The hormones produced by the adrenal cortex and by the gonads (ovaries and testes) are also steroids.

3. *Phospholipids*

These are complex substances whose functions are as yet poorly understood. As their name suggests they contain phosphate groups. They seem to be important in the transport of fats in the blood, in the manufacture of cell membranes and in the structure of nervous tissue.

PROTEINS

Proteins consist mainly of carbon, hydrogen, oxygen and nitrogen with much smaller quantities of sulphur. The large protein molecules are built up from smaller substances known as amino acids, of which about twenty types are important in human biochemistry. The amino acids are substances which carry both carboxylic (COOH) and amino (NH$_2$) groups.

$$
\begin{array}{c}
CH_3 \\
| \\
H_2N-C-COOH \quad \text{Alanine} \\
| \\
H
\end{array}
$$

(1) H$_2$N–⬡–COOH   H$_2$N–⬡–COOH   H$_2$N–⬡–COOH

(2) H$_2$N–⬡–CO-HN–⬡–CO-HN–⬡–COOH
           +H$_2$O   +H$_2$O

Figure 3.4 Alanine is a typical amino acid. (1) Shows three separate amino acids. (2) Shows them linked to form a peptide chain with the elimination of water.

Amino acids can link together in chains, with the carboxylic acid group of one amino acid joined to the amino group of the next amino acid. Such chains are known as peptides and the links between the amino acids are known as peptide bonds. Proteins may contain one, two, three or four peptide chains. In the full protein molecule, the chains are folded and coiled up upon themselves in complex ways.

The proteins in the body are important in two main ways. They make up parts of the structure in many organs (e.g. the tendons of muscles). They also act as enzymes which are important in virtually all biochemical reactions.

NUCLEIC ACIDS

These have been much studied in recent years. They are important because the genetic material in the chromosomes consists of nucleic acids. Similarly, the ribosomes, structures in the cytoplasm of cells where most protein synthesis seems to take place, are rich in nucleic acids. The nucleic acids are the vital materials which hand on inherited characters from generation to generation. They

are also important in directing the day to day life of each cell: disorders of the control which they normally exert over the cell's behaviour may be important in the causation of cancer.

Nucleic acids too are made up of simpler building blocks. Each molecule contains sugars, phosphate groups and complex organic materials known as bases. The nucleic acids are divided into two great groups depending on whether they contain the pentose sugar, ribose, or the slightly different pentose, deoxyribose. The former are known as ribonucleic acids (RNA's) and the latter as deoxyribonucleic acids (DNA's). The bases are shown in table 3.1. DNA is found primarily in cell nuclei, while

Table 3.1 The bases found in the nucleic acids.

DNA	RNA
Adenine (A)	Adenine (A)
Thymine (T)	Uracil (U)
Guanine (G)	Guanine (G)
Cytosine (C)	Cytosine (C)

RNA is found primarily in the cytoplasm. Both types of nucleic acid consist of chains of pentose molecules joined to one another by phosphate groups. The bases are attached to the sugars. In DNA two such chains are wound around one another in a helical formation (the famous 'double helix').

ENZYMES

Enzymes are the protein molecules on which the whole of metabolism depends. Almost all the chemical reactions which occur in the body could take place without enzymes but they would take place far too slowly to have any biological value. Enzymes speed up and direct virtually all the chemical reactions of metabolism. The main characteristics of the enzymes themselves are:

1. They are all protein in nature but many of them have some other substance such as a metallic ion attached to the protein molecule.

Ribonucleic acid

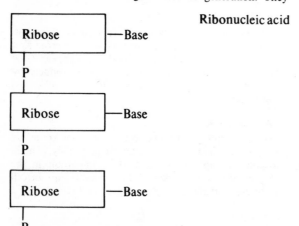

Figure 3.5 The constitution of RNA.

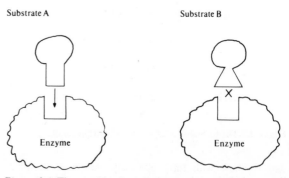

Figure 3.6 The combination of enzyme with substrate.

2. Most enzymes are specialised to carry out a particular function.

3. All enzymes require special environmental conditions if they are to function normally.

The specialisation of function occurs because in order to work, an enzyme must actually combine with the substance which it is chemically changing. Substances acted upon by enzymes are known as substrates. A combination between an enzyme and its substrate may be likened to a lock and key mechanism. The enzyme is the lock: because it is of a particular chemical structure, only substrates whose chemical structure can fit the lock can become attached to the enzyme. Only when the substrate has become attached to the enzyme can the enzyme function. Enzymes function in three main ways:

1. They may break the substrate molecule into two parts. This happens particularly in digestion.

2. They may break off a small part of the substrate molecule, usually a hydrogen atom or a CH_3 group, or a water molecule or a carbon dioxide molecule.

3. They may build up the substrate molecule by adding a small part to it.

Whatever happens, once the change has been brought about, the substrate is released from combination with the enzyme and the enzyme is ready to repeat the process on another substrate molecule. From this brief description the following important points about enzyme behaviour may be deduced:

1. Most enzymes are highly specialised so that they can react with only one type of substrate.

2. Each enzyme carries out only one process. Therefore in order to break down a large molecule, or in order to build up a large one from much smaller units, many enzymes are required, each performing one stage in the whole process. The enzymes do not swim around inside cells in a form of soup. Instead they are attached in sequence to membranes. When a substrate molecule has been dealt with by one enzyme, the substrate is passed immediately to the enzyme which carries out the next stage in the process and which is attached to the membrane immediately adjacent to the first enzyme.

3. There is a limit to the rate at which a single enzyme molecule can work. If there is an excess of substrate available then all the enzyme molecules will be working at maximum capacity and the more enzyme molecules there are the faster will the substrate be changed. If all the enzyme molecules are not working maximally, then the rate of change of the substrate depends on how rapidly the substrate can be supplied to the enzymes.

4. Many enzymes act be either adding to or removing from substrates, atoms or small groups of atoms. If the former type are to work effectively they must be constantly supplied with the atoms and small groups which they use. If the latter are to function properly the atoms and small groups must be continuously removed or they will accumulate and gum up the works. The supply and removal of these atoms and small groups is the work of substances known as coenzymes. Many of the vitamins, particularly those of the B group, appear to be so important because they function as coenzymes in vital reactions.

Protein molecules are very easily damaged and if they are to function properly, enzymes require very special conditions. The temperature must be kept within very narrow limits: if it rises too high or falls too low, the rate at which the enzymes work is altered and the enzyme molecules themselves may be destroyed (or denatured as the process is sometimes known). Most enzymes in mammals work best at about the temperature of the blood and this is a major reason why the temperature of the blood and of the body must be kept constant. The pH of the body fluids must also be kept more or less constant. Most enzymes again work best at the pH of the blood. There are some exceptions to this rule, however, for example: the enzymes in the stomach work best in quite strong acid and cease to function at the pH of the blood. Finally the osmotic pressure and chemical composition of the body fluids must also be kept constant if the enzymes are to work normally. This maintenance of a constant chemical composition is primarily the job of the kidney.

Enzymes are now named by adding '-ase' to the name of the substrate on which the enzyme acts. For example, enzymes which break down proteins are called proteases, those which break down lipids are called lipases and so on. Some of the enzymes which were discovered many years ago were named before this system was introduced. These enzymes usually keep their own names, like the pepsin which digests proteins in the stomach and the trypsin of pancreatic juice which digests proteins in the intestine.

OUTLINE OF CATABOLISM

Carbohydrates, fats and proteins are all capable of being broken down to supply energy. Although these substances are so different, the final and most important stages of their destruction are shared with one another. The first aim of catabolism is to break down, usually without the aid of oxygen, the large complex molecules into smaller ones. The small molecules resulting from the breakdown of all three foodstuffs can then enter the series of reactions known as the citric acid, Krebs, or tricarboxylic acid cycle. Once in the cycle they are further broken down with the aid of oxygen to give water, carbon dioxide and large amounts of ATP.

CARBOHYDRATE METABOLISM

Carbohydrates occur in human food in three particularly important forms, as the polysaccharide starch, and as the disaccharides sucrose (cane sugar) and lactose (milk sugar). All three must first be broken down to the monosaccharides glucose, fructose and galactose before they

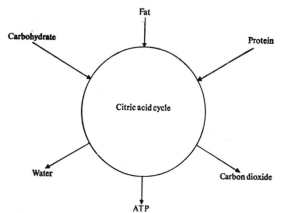

Figure 3.7 The central position of the citric acid cycle in metabolism.

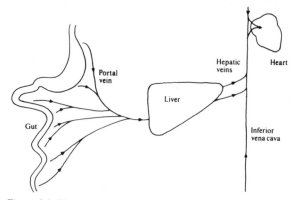

Figure 3.8 The arrangement of the portal vein and the liver.

can be absorbed from the gut into the blood. This process is known as digestion and the main stages are:

1. Saliva contains an enzyme known as amylase or ptyalin, which can break down starch into smaller units. The process does not continue very far as amylase is inactivated by the acid gastric juice.

2. Gastric juice contains no enzyme which breaks down carbohydrates. However, the acid itself may help to break up the tough starch granules, so making them more easily digested in the intestine.

3. The pancreatic juice which is poured into the duodenum also contains amylase which completes the breakdown of starch to the disaccharide maltose. The bicarbonate in the pancreatic juice neutralises the gastric acid and so allows the enzyme to work.

4. The wall of the intestine itself manufactures enzymes which break down the disaccharides to monosaccharides. Maltase breaks down maltose to glucose. Lactase breaks down lactose to glucose and galactose. Sucrase (sometimes known as invertase) breaks down sucrose to glucose and fructose. If one of these enzymes is absent at birth, the infant will be unable to digest one of the disaccharides (most commonly lactose). The undigested food passes through the gut causing diarrhoea: ultimately malnutrition will occur because the baby cannot obtain enough calories. A remarkable improvement occurs once the offending disaccharide is removed from the diet.

THE PORTAL VEIN AND THE LIVER

The monosaccharides, glucose, galactose and fructose, pass through the intestinal wall into the blood of the hepatic portal system. The hepatic portal vein carries all the blood from the gut to the liver. Only after passing

through the liver does the blood from the gut mix with the blood from the rest of the body returning to the right side of the heart. In the liver, much of the glucose is removed from the blood and is built up into the polysaccharide, glycogen, in which form it is stored. Much of the galactose and fructose is also removed from the blood and liver enzyme systems convert these substances into glucose, the central substance in carbohydrate metabolism. These converting enzyme systems may also be congenitally defective. If they are, either galactose or fructose may not be converted to glucose. A deficiency of the galactose enzymes is not uncommon and is particularly dangerous because of the large amounts of galactose formed from the lactose of milk. Galactose accumulates in the blood in large amounts (galactosaemia) and may damage the brain. Once the condition is recognised and lactose is removed from the diet, a rapid improvement occurs.

Between meals, when no glucose is entering the blood from the gut, the glycogen in the liver is slowly broken down. The glucose produced in this way is released into the blood in order to keep the level of blood glucose constant. It is important that glucose concentrations in the blood should be kept constant because glucose can be used as an energy source by virtually every cell in the body. The constancy is particularly vital for the function of the brain. Most other cells in the body can get their energy from carbohydrate or fat but this does not seem to apply to the nerve cells. The brain seems to rely entirely on glucose for its energy supply and it is incapable of using fat or protein. Therefore if the blood glucose concentration becomes too low, the brain cells may be permanently damaged.

WHAT HAPPENS TO BLOOD GLUCOSE?

Most of the glucose in the blood is removed from it and immediately used by cells to provide energy. Some is taken up by the muscles and heart for storage in the form of glycogen in order to tide over possible emergencies. Some is taken up by fatty tissue and converted into fat, another form of food storage.

The direct breakdown of glucose by cells can be divided into two stages, anaerobic in which no oxygen is required, and aerobic in which oxygen is essential. In the

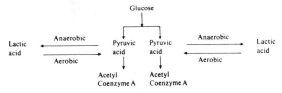

Figure 3.9 The breakdown of glucose to acetyl CoA.

first of these states, glucose is broken down to pyruvic acid: in the absence of oxygen, the process can go no further and the pyruvic acid is converted to lactic acid, a blind alley in metabolism. In humans and other mammals oxygen is usually available and lactic acid is formed primarily during muscular exercise when the oxygen supplies reaching the muscle may not be quite sufficient. However, many micro-organisms, such as yeasts, may not have the enzymes to carry the breakdown of pyruvic acid further. They, like humans, convert glucose to pyruvic acid but then they convert the pyruvic acid to ethyl alcohol: this reaction is the foundation of the whole brewing and wine making industry.

In exercising humans, some of the lactic acid produced stays in the muscles while some escapes into the blood and reaches the liver. When the exercise stops and more oxygen is again available, the lactic acid is converted back to pyruvic acid which is either oxidised completely or converted back to glucose and glycogen. Up to the state of pyruvic acid, only very small amounts of ATP are produced from each glucose molecule.

AEROBIC METABOLISM

When ample oxygen is available, the pyruvic acid is converted into a very important substance known as acetyl coenzyme A. This conversion is particularly interesting because it demonstrates the part which vitamins

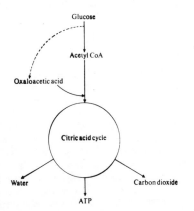

Figure 3.10 Oxaloacetic acid and the entry of acetyl CoA into the citric acid cycle.

can play in the body. Four vitamins, thiamine, lipoic acid, pantothenic acid and nicotinic acid are essential for the reaction to occur.

Acetyl coenzyme A is important because it is the means whereby carbohydrates can enter the citric acid cycle which completely oxidises them to give water, carbon dioxide and large amounts of ATP. The conversion of one glucose molecule to pyruvic acid yields only four molecules of ATP, while the conversion of the two molecules of acetyl coenzyme A formed from one glucose molecule to carbon dioxide and water yields thirty-four molecules of ATP. It is obvious that aerobic metabolism is a much more effective and important process.

Finally, as will become apparent later when discussing diabetes, it is important to note that acetyl coenzyme A enters the citric acid cycle by combining with a substance known as oxaloacetic acid. Oxaloacetic acid too can be formed from glucose via a side path of metabolism.

FAT METABOLISM

Fats are just as important as carbohydrates in the supply of the energy needs of the body. Most of the fat in the food is in the form of triglycerides. Fats differ from carbohydrates in that although they must be partially digested, they do not need to be broken down completely into their basic building blocks of glycerol and fatty acids before they can be absorbed across the gut wall.

No digestion of fat takes place in the saliva and little if any in the stomach. However, in the stomach the fat is softened and possibly small amounts may be broken down by the direct action of the stomach acid. But there can be no doubt that the major part of fat digestion begins when the food leaves the stomach and enters the first part of the small intestine, the duodenum. There it meets with two important juices, the bile secreted by the liver and the pancreatic juice secreted by the pancreas.

The bile does not actually digest the fat. It contains no fat-breaking enzymes (lipases) but it does contain the bile salts of glycocholic and taurocholic acids. The bile salts are detergents, similar in many ways to the detergents which are used for washing up dirty dishes. The first problem of fat digestion is how can the lipases which are found in the pancreatic juice get at the fat? The enzymes are secreted in the form of a watery solution which cannot penetrate into the fat globules of the food just as ordinary water cannot break up the large fat globules in the frying pan. The enzymes can get at the fat molecules only if the large, unwieldy fat globules are broken down into tiny droplets. This breaking up of the globules is performed by means of the detergents in the bile and is similar in principle to the clearing of the frying pan fat by washing-up fluid. The process is known as emulsification. Once the fat is in the form of tiny droplets, then it is much more accessible to the enzymes of the pancreatic juice.

The pancreatic lipases break off fatty acids from the triglycerides. The droplets become still smaller and consist of a complex mixture of glycerol, free fatty acids, monoglycerides, diglycerides, triglycerides and bile salts. These droplets can then pass through the gut wall even without the complete digestion of triglycerides to give glycerol and free fatty acids. Although the droplets can pass through the gut wall they are still too large to cross the walls of the capillaries to enter the blood. Instead they enter the much more porous lymphatic vessels. They are carried by the lymph directly into the venous blood, by-passing the portal system and the liver.

FAT CATABOLISM

There are three main stages in the breakdown of fats within the body. First of all the glycerides are split up to give glycerol and free fatty acids. Glycerol is a substance which is part way between glucose and pyruvic acid in structure and so it can easily be broken down by the enzymes which metabolise glucose. Fatty acids are quite different. Two carbon atoms at a time are split off from the chain of carbon atoms in the fatty acid molecule. These two-carbon fragments are converted to acetyl coenzyme A. Finally the acetyl coenzyme A combines with oxaloacetic acid and is completely oxidised by the citric acid cycle. Thus both fats and carbohydrates are finally broken down by the same route.

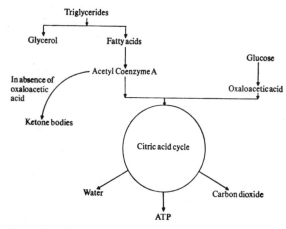

Figure 3.11 The breakdown of fats.

It is worth noting that both fats and carbohydrates must enter the citric acid cycle by combining with oxalo-acetic acid. Oxaloacetic acid is manufactured from carbohydrate. This means that carbohydrates can supply their own oxaloacetic acid and do not need any fat oxidation for their complete breakdown. In contrast the fats cannot supply their own oxaloacetic acid and hence they cannot be fully oxidised unless some carbohydrate is also available.

A deficiency of carbohydrate is most likely to occur in two situations. In starvation, the carbohydrates are rapidly used up and the body must rely on its fat stores for its energy supply. In diabetes mellitus there may be plenty of carbohydrate available but it cannot be used

normally because of lack of insulin. In these circumstances, because of lack of oxaloacetic acid, fat breakdown tends to be halted at the stage of acetyl co-enzyme A. Acetyl coenzyme A molecules therefore accumulate and react with one another to give three other substances, acetoacetic acid, betahydroxybutyric acid and acetone. These three substances are known as ketone bodies and when they accumulate, the patient is said to be suffering from ketosis.

FAT STORAGE

Carbohydrate is stored in small quantities in liver and muscles in the form of glycogen. These glycogen stores can supply energy needs for only 24–48 hours after which they are completely used up. The main energy stores in the body are in the form of fat. This seems to be because a given weight of fat yields considerably more energy that the same weight of carbohydrate. It is therefore more efficient to store energy in the form of fat.

The fat is stored in special cells in the tissue known as adipose tissue. Adipose tissue is mainly found immediately beneath the skin, but there are often large amounts in the peritoneal cavity in the folds of the omentum. Under the skin, the breasts, buttocks and abdominal wall seem to be the sites most preferred. The fat in adipose tissue is found in the form of triglycerides. These may be built up directly from the fatty acids in the blood or they may be made indirectly from carbohydrate in the food not immediately required by the body. When a person takes in more calories than he or she needs, the excess energy is stored in the form of fat. When more calories are used up than are supplied in the food, the extra calories needed are provided first by the breakdown of glycogen and then by the breakdown of fat.

ESSENTIAL FATTY ACIDS

Fats are obviously of major importance in the supply of energy. But some types of fat have other roles as well. The steroids are hormones and the phospholipids are essential for the structure of cell membranes and of the central nervous system. The unsaturated essential fatty acids play a poorly understood role but one which may turn out to be of the utmost significance. The important essential fatty acids are called linoleic, linolenic and arachidonic acids. They are similar to vitamins in that they are required in very small quantities in the diet because they cannot be manufactured by the body from other substances. Their precise actions are uncertain but they appear to be essential for the manufacture of normal cell membranes. If experimental animals are completely deprived of essential fatty acids they develop a syndrome in which the skin and kidneys are damaged and reproduction fails. However, they are so widely distributed in the diet that these extreme conditions resulting from total lack are unlikely to occur naturally. Their main interest is that there is some evidence that a partial deficiency in humans may produce defective arterial walls and lead to the development of the very common disease, athero-sclerosis, but this is as yet unproven.

PROTEINS

The provision of energy is the major function of both the fat and carbohydrate in the food. Both fat and carbohydrate contain only carbon, hydrogen and oxygen and can therefore be completely broken down to give carbon dioxide and water. Proteins differ in two main ways. First, their major function is to provide the amino acids from which enzymes and other important constituents of the body can be manufactured: they can be broken down to supply energy but this is a function of secondary importance. Secondly, in addition to carbon, hydrogen and oxygen they contain large amounts of nitrogen and small amounts of other substances such as sulphur. They cannot therefore be broken down to carbon dioxide and water alone since neither carbon dioxide nor water contains nitrogen. The nitrogen must be disposed of somehow. This is primarily the task of the kidneys which excrete most of it in the form of urea with smaller amounts as uric acid and ammonium ions.

DIGESTION

The proteins taken in with the food cannot be used directly by the body. Before they can be absorbed into the blood they must be broken down in the gut to give their constituent amino acids.

The first important stage of protein digestion takes place in the stomach. There an inactive enzyme, pepsinogen, is secreted by glands in the stomach wall. It is

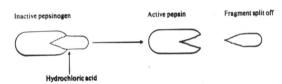

Figure 3.12 The activation of pepsin.

important that this protein-digesting enzyme should be secreted in an inactive form, otherwise it would digest the cells in which it was made. On entering the stomach it comes into contact with the acid gastric juice which is secreted by other cells in the stomach wall. The acid breaks off a small part of the inactive pepsinogen molecule and converts it to active pepsin. The pepsin can then begin to break up the amino acid chains of which the protein molecules are made. It works best at a pH of 2–3 which is the highly acid pH normally found in the stomach.

The digestion of proteins to the level of single amino acid units is completed in the intestine. Protein-splitting enzymes are found both in the juice secreted by the pancreas and in the secretions of glands in the wall of the small intestine itself. All these enzymes work best when pH is nearly neutral (around 7). The first stage therefore in protein digestion in the intestine is the neutralisation of the acid material coming from the stomach by the large amounts of bicarbonate found in the pancreatic juice.

The pancreas secretes a number of inactive enzyme precursors of which the best known is trypsinogen. Again these substances must be made in an inactive form so that they do not digest the pancreas itself. In the disease pancreatitis, the enzymes are abnormally activated within the pancreas itself. They therefore digest the gland and escape into the abdominal cavity doing a great deal of damage and causing intense pain. Under normal circumstances, however, trypsinogen is not converted to active trypsin until it leaves the pancreatic duct and enters the small intestine. There it meets a substance called enterokinase which is secreted by the intestine wall. Enterokinase splits off a small part of the trypsinogen molecule leaving the active enzyme trypsin. By means of both pancreatic and intestinal enzymes, the proteins are broken down into the amino acids which can cross the gut wall and enter the portal blood.

ESSENTIAL AMINO ACIDS

Twenty-one amino acids have so far been discovered in the proteins found in mammals. Some of these cannot be manufactured in the human body in sufficient amounts to supply the body's needs. They must therefore be supplied in the food and are known as essential amino acids.

Table 3.2 Essential and non-essential amino acids.

Essential	Non-essential
Arginine	Alanine
Histidine	Aspartic acid
Leucine	Cystine
Isoleucine	Glutamic acid
Lysine	Glycine
Methionine	Hydroxyproline
Phenylalanine	Proline
Threonine	Serine
Tryptophan	Tyrosine
Valine	Cysteine
	Hydroxylysine

Protein is sometimes said to be of high biological value when it contains all the essential amino acids and of low biological value when one or more of the essential acids are missing.

The other amino acids are sometimes called non-essential simply because they do not need to be supplied in the food: they can be manufactured by the body itself. Surplus quantities of the amino acids the body does not immediately need can be used to manufacture those amino acids which it does need.

FATES OF AMINO ACIDS

Four types of thing may happen to the amino acids which result from the digestion of protein in the food.

1. They may be used unchanged for the manufacture of new protein.

2. If there is an excess of a particular amino acid, that acid may be converted to another amino acid of which there is a shortage.

3. The amino acids may be broken down. The parts containing carbon, hydrogen and oxygen enter the citric acid cycle and are oxidised to supply energy. Thus the citric acid cycle is the final route of oxidation for all three major foodstuffs, fats, carbohydrates and proteins. The nitrogen-containing part of the amino acid molecule cannot be dealt with in this way. Instead it is usually converted to urea and excreted in the urine.

4. They may be used for the manufacture of non-protein nitrogen containing substances such as thyroid hormone, the nucleic acids and the haem part of haemoglobin.

THE FORMATION OF UREA

The first stage in the breakdown of the amino acids is the splitting off of the amino groups to give ammonia. This can occur in many organs but is most likely to take place in the liver. The process is known as deamination.

Ammonia is a highly toxic substance and if allowed to accumulate it may damage many organs and in particular the brain. It must therefore be converted as quickly as possible to something which is relatively harmless, highly soluble in water and excretable in the urine. The substance with these properties is urea and the conversion of ammonia to urea takes place in the liver. When ammonia is formed in other organs the blood rapidly takes it to the liver where it is rendered harmless. The sequence of reactions in the liver by which ammonia is converted to urea is known as the ornithine cycle.

NUCLEIC ACIDS AND PROTEIN SYNTHESIS

The chromosomes found in the nucleus of every cell are responsible for the development of that cell and for the control of the behaviour of the mature cell. The chromosomes seem to consist primarily of the nucleic acid known as deoxyribonucleic acid or DNA. The DNA carries the 'plans' for the structure of every protein molecule manufactured by the cells. Since the proteins include the enzymes and since the enzymes are responsible for the manufacture of all non-protein substances in a cell, the DNA is in the end responsible for the manufacture of all substances made by each cell.

CELL DEVELOPMENT

The most important constituents of the human fertilised egg are the forty-six chromosomes, twenty-three derived from the father's sperm and twenty-three from the mother's ovum. These forty-six chromosomes carry the plans for the whole development of the body. Whenever a cell divides, exact copies of the forty-six chromosomes are made so that each of the pair of daughter cells contains a set of chromosomes identical to the set of the parent cell.

Several thousand different types of proteins are probably required for the normal working of the human body. Most of these are enzymes. Some are hormones such as insulin and growth hormone and others are substances serving a quite different function such as the globin part of the blood pigment haemoglobin. The DNA

in the forty-six chromosomes carries the plans for the manufacture of all these proteins. The part of a chromosome which deals with the manufacture of one protein is known as a gene. It is possible for one gene and therefore one protein to be faulty even though the rest of the body is normal. For example, in haemophilia the gene which carries the instructions for the manufacture of the protein antihaemophilic globulin (a clotting factor) is faulty. This means that the globulin cannot be manufactured normally and the blood does not clot normally. In galactosaemia the gene which carries the instructions for one of the enzymes involved in the conversion of galactose to glucose is faulty and therefore galactose accumulates in the blood. Such genetic errors which affect the biochemistry of the body are sometimes known as inborn errors of metabolism.

Although the DNA in every cell in the body of any individual is identical and is therefore capable of stimulating the manufacture of every protein that the body can make, in practice each cell becomes specialised so that it manufactures only a few types of protein. Only a few genes are allowed to function and the activity of the others is more or less permanently suppressed. For example, red cells are specialised to manufacture haemoglobin: the parts of the DNA which carry the 'plans' for insulin or for the enzymes used in the synthesis of fats are inactive. In contrast, the cells in the islets of Langerhans in the pancreas produce primarily insulin while the cells in adipose tissue produce primarily those enzymes required to deal with fats. Each cell is therefore specialised so that it produces only a limited number of the proteins which it could theoretically manufacture.

In addition, the manufacture of the proteins which each cell does produce is also strictly controlled. When a

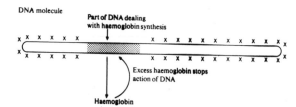

Figure 3.13 The action of DNA in red cell formation.

cell contains sufficient of a particular protein, information about this is sent back to the nucleus; the DNA stops further production of that protein. When the amount of that protein in the cell falls again, the DNA is again informed and this time it orders the manufacture of more of that protein. In this way the concentration of each type of protein within a cell is kept close to the desired level.

THE GENETIC CODE

Each protein molecule consists primarily of a chain of amino acids. The properties of the protein molecule depend on the types of amino acids in the chain and on the order in which those amino acids occur. The DNA must therefore contain the plans for each protein molecule in terms of plans for a chain of amino acids. How does it do this?

Earlier we saw that DNA contains four types of bases, adenine (A), thymine (T), guanine (G) and cytosine (C). It has been found that the sequence in which these bases occur on the DNA molecule determines the sequence in which the amino acids occur in the protein molecule. It has also been found that in order to specify one amino acid a sequence of three bases is required—the sequence of bases which specifies each amino acid is sometimes known as the genetic code. This may perhaps be best

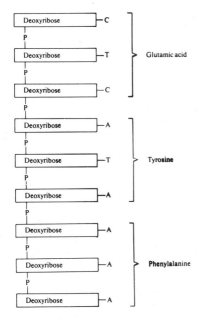

Figure 3.14 Schematic diagram of a short length of a DNA molecule showing how the genetic code works.

understood if specific examples are used. The amino acid phenylalanine is specified by the base code AAA, glutamic acid is specified by the base code CTC, while tyrosine is specified by the base code ATA. Thus the sequence of bases CTC–ATA–AAA on DNA would be the plan for a short peptide chain of composition glutamic acid-tyrosine-phenylalanine. Longer sequences of bases on DNA can specify the amino acid composition of any protein in the body.

PROTEIN SYNTHESIS

The mechanism of protein synthesis has been worked out only recently. The bulk of such synthesis takes place in structures found in the cytoplasm of cells known as ribosomes. The ribosomes are largely made up of a special type of ribonucleic acid known as ribosomal RNA. The ribosomes are obviously well away from the DNA which

is the nucleus. A big problem in protein synthesis is therefore how the DNA gets its message across to the ribosomes. It may be helpful to consider the building of a protein molecule as similar in many ways to the building of a house. In house building, four categories of men are required:
1. An architect to draw up the plans.
2. A foreman to look at the plans and to supervise the actual construction of the house.
3. A gang of labourers to bring the necessary building materials to the site.
4. Another gang of labourers actually to put up the house.

In building a protein molecule the plans are provided by the DNA of the cell nucleus: it acts as the architect. The protein molecules are assembled from the amino acids by the ribosomes: they are the labourers who actually put up the building. The amino acids are brought to the ribosomes by another special type of RNA known as soluble or transfer RNA: the molecules of transfer RNA appears to float free in the cytoplasm. There are in fact over twenty different types of transfer RNA, each one specialised to combine with a particular amino acid. For example, there is a transfer RNA which combines with leucine, another which combines with alanine, another which combines with threonine and so on. The job of these transfer RNA molecules is to pick up the amino acids which reach the cell from the blood and to carry them to the ribosomes. There the link between the transfer RNA and the amino acid is broken. The amino acid is used for protein synthesis and the transfer RNA is free to go off to collect another amino acid molecule.

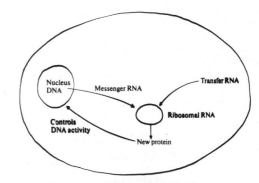

Figure 3.15 The synthesis of proteins in a cell.

The remaining problem therefore is the link between nuclear DNA and ribosomal RNA. What acts as the 'foreman' to instruct the ribosomes how to carry out the plans of the nucleus? The link is in fact made by yet another type of RNA, this time known appropriately enough as messenger RNA. Suppose that the nucleus decides that a particular protein needs to be manufactured. That protein consists of a chain of amino acids and the plan for it is represented by a chain of groups of three bases on the DNA molecule. The first thing that happens is that an impression of this length of DNA is made by manufacturing a messenger RNA molecule which is complementary to it. This is possible because each base in the DNA molecule orders the insertion of another base in the messenger RNA molecule. DNA adenine orders RNA uracil, DNA cytosine orders RNA guanine, DNA thymine orders RNA adenine. Thus a sequence of nine DNA bases, CTC–ATA–AAA, would become on the messenger RNA molecule, GAG–UAU–UUU. Once the full messenger RNA molecule has been made in this way it leaves the nucleus and moves out to a ribosome in the cytoplasm. The ribosomal RNA seems to be capable of 'reading' the code. For example, it 'sees' that the first three bases are GAG and it 'knows' that this sequence means glutamic acid. It therefore starts off the protein molecule by taking a molecule of glutamic acid from the transfer RNA. It then 'reads' UAU and so a molecule of tyrosine is taken from the cytoplasm and tagged on to the glutamic acid. UUU means phenylalanine and so the third amino acid is phenylalanine. The ribosome thus reads the whole chain of the messenger RNA molecule and builds up a corresponding chain of amino acids. On being completed the amino acid chain is released and spontaneously folds up to become the complete protein molecule.

SIGNIFICANCE IN DISEASE

We now know that most viruses consist of an outer protein coat with an inner core of nucleic acid. Most of the viruses which attack humans contain DNA although some contain RNA. The virus becomes attached to a cell by means of its protein coat. The protein then seems to bore a hole in the side of the cell and injects the nucleic acid into the cell interior. Once inside the nucleic acid can do two things. First it may manufacture its own equivalent of messenger RNA. This goes to the ribosomes which are then instructed to manufacture new virus particles instead of the normal cell proteins. The ribosomes are thus partially taken over by the invading virus. Secondly the virus seems to interfere with the normal controls over protein manufacture. When a particular

protein accumulates in the cell, the DNA in the nucleus is no longer informed and so production of that protein continues and is not stopped as it usually would be. The cell must grow to accommodate the extra protein and quite soon it divides. The virus thus promotes abnormal cell multiplication. In some cases the virus seems to have primarily the first action. It takes over the ribosomes so completely that the cell ceases to manufacture its own protein and so soon dies. This seems to happen with a disease like smallpox. On the other hand, if the virus allows the cell to manufacture some of the cell's own protein, the cell will not die but will grow, divide and multiply in an uncontrolled way. This is what happens with the familiar warts which are caused by a virus. More seriously it occurs much more dramatically in the large numbers of tumours in animals which are now known to be caused by viruses. No human tumour has yet been shown to be virus-induced but most research workers now believe that some types of human cancer at least are caused by viral infection.

In all types of cancer, the normal controls which stop excess protein synthesis and cell multiplication appear to be lost. As the cell manufactures protein and that protein accumulates, the process is no longer stopped by information carried back to the DNA. The cell must divide and rapid multiplication takes place. Furthermore, many of the controls over the behaviour of DNA which develop during early growth seem to be lost. Instead of being specialised to produce just a few types of protein, the rapidly dividing cells start making substances they do not normally manufacture. The best known examples of this phenomenon are tumours of the lung which not uncommonly suddenly start manufacturing hormones which are normally produced only by the specialised cells of the pituitary gland. ACTH and ADH are frequently manufactured. The ACTH stimulates the adrenal cortex to abnormal activity and so the patient with lung cancer may present with Cushing's syndrome. The ADH prevents normal water excretion and so the patient becomes waterlogged.

The study of nucleic acids and of protein synthesis is thus not simply an academic problem. It contains the seeds of the answers to cancer and to viral disease.

THE LIVER

As far as biochemistry is concerned, the liver is by far the most important organ in the body. A list of the things which it does, which are mentioned elsewhere in the book, is given in table 3.3. This chapter deals with some of those liver functions which are not dealt with elsewhere.

STRUCTURE

The liver consists of columns of cells all of which appear to be identical in structure and function. Serving these cells, by carrying to them or from them fluids of various sorts, are five separate systems of tubes.

1. *Portal Venous System*

The portal vein collects blood from the gut which after a meal is rich in digested food. On entering the liver, the

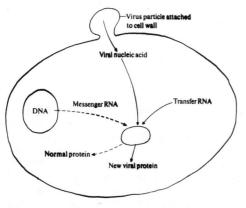

Figure 3.16 The action of a virus on cell protein synthesis.

Table 3.3 The main biochemical functions of the liver not discussed in this section.

Protein metabolism
1. Deamination of proteins, and manufacture of urea from ammonia.
2. Synthesis of plasma proteins.
3. Synthesis of non-essential amino acids.

Carbohydrate metabolism
1. Synthesis and storage of glycogen and regulation of blood glucose.
2. Oxidation of carbohydrate.
3. Conversion of carbohydrate to fat.

Lipid metabolism
1. Synthesis of fatty acids and phospholipids.
2. Oxidation of fats.

Nucleic acid metabolism
1. Synthesis of the bases found in the nucleic acids.

---

portal vein breaks up into smaller and smaller vessels eventually reaching flabby, thin-walled structures known as sinusoids. These allow the blood to come into contact with the liver cells. The sinusoids contain the so-called reticulo-endothelial cells in their walls. The reticulo-endothelial cells are phagocytic: that is to say that they can engulf solid particles from the blood. Similar cells, part of the reticulo-endothelial system, are found in sinusoids in the bone marrow and spleen. The reticulo-endothelial cells in the liver are particularly important in the removal of old red blood cells and of bacteria which sometimes enter the portal blood from the gut.

2. *Hepatic arteries*
The total flow of blood through the liver after a meal is in the region of 1500 ml/min. About four-fifths of this comes from the portal vein and is partly deoxygenated as it has already passed through the gut. In contrast, the hepatic arteries supply about 300 ml/min of freshly oxygenated blood direct from the aorta. The hepatic arteries too send their blood to the liver sinusoids.

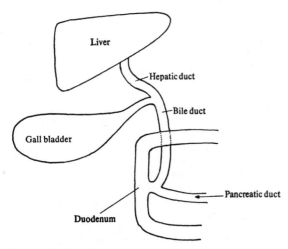

Figure 3.17 The bile system.

3. *Hepatic veins*
These collect the blood from the sinusoids and pour it into the inferior vena cava which returns it to the right side of the heart.

4. *Biliary Tract*
This is a blind-ended system of tubes which receives the substances secreted by the liver cells. It starts as very tiny tubes in close contact with the liver cells. These tubes join together eventually, forming the important ducts shown in fig. 3.17. The bile is emptied into the duodenum. The gall bladder is a blind sac whose main function is bile storage. It does not secrete bile itself but it does remove some of the water from bile so making it more concentrated.

5. *Lymphatics*
These collect any excess intercellular fluid and return it to the blood.

COMPOSITION OF BILE
Bile is a substance whose importance lies partly in its role in the digestion of fats and partly in the fact that it is the route by which many substances are excreted from the body. The bile salts (sodium salts of glycocholic and taurocholic acids) act as detergents and help in the digestion of fats. The important substances excreted in the bile are:

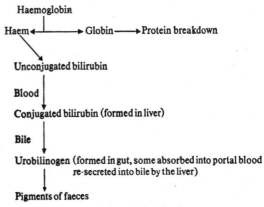

Figure 3.18 The metabolism of the bile pigments.

1. *The bile pigments*. These are mainly formed by the breakdown of the haem part of the blood pigment haemoglobin. The brownish-yellow bilirubin is formed first. When stored in the gall bladder bilirubin may be converted to the dark green biliverdin. The bile pigments are responsible for the normal brown colour of the faeces and will be further discussed later in the chapter.

2. *Cholesterol*. This steroid is secreted in large amounts by the bile.

3. *Alkaline phosphatase*. This is an enzyme which is made in the bones and which is removed from the blood

by the liver. If the liver is damaged or if bile excretion is obstructed, the concentration of alkaline phosphatase in the blood rises.

4. *Penicillin and ampicillin.* Because it is excreted in the bile, ampicillin is often used for biliary tract infections.

5. *Various iodine-containing compounds.* These are artificially manufactured substances which are opaque to X-rays: this means that they can be identified in the body by the shadows they throw on X-ray films. Some of the substances used are very opaque indeed and when injected intravenously they outline first the biliary tree and then the gall bladder. Others, usually given by mouth are not so opaque. They cannot be 'seen' in ordinary bile when it is first secreted and so they cannot be used to outline the biliary tree. However, when the bile is concentrated in the gall bladder they can be identified and so can be used to outline the bladder and to assess its ability to concentrate the bile.

THE GALL BLADDER
This does not secrete bile. It merely stores it and removes some water from it so making it more concentrated. Concentrated solutions tend to deposit substances dissolved in them in the form of solid material. In the gall bladder, such solid deposits make up the familiar gall stones. The major constituents of a gall stone are cholesterol and the bile pigments. These substances allow X-rays to pass through them and so they cannot be seen on a plain X-ray. Only the 10–15% of gall stones which contain moderate amounts of calcium can be seen in this way. The stones which are not radio-opaque can be seen by giving one of the iodine compounds concentrated in the gall bladder. These normally fill the bladder producing a smooth, solid shadow. If many gall stones are present, the bladder will not be filled evenly and the shadow will appear to have many 'holes' in it.

The factors which make some people form many gall stones are as yet poorly understood. They tend to be commonest in fat, middle-aged ladies who have borne many children, but no one can be said to be exempt. Infections of the biliary tract seem in some cases to be associated with gall stones.

OTHER FUNCTIONS OF THE LIVER
The most important functions of the liver not yet mentioned are:

1. *Storage.* The liver stores glycogen, iron and vitamins A, D, and B_{12}.

2. *Red cell destruction.* The reticulo-endothelial cells in the sinusoids are an important part of the system which removes old red cells from the blood and destroys them.

3. *Red cell manufacture.* During fetal development before birth, the liver is an important site of red cell manufacture. By the time of birth, however, this function has been completely taken over by the bone marrow. Rarely, if the bone marrow is destroyed, the liver may take up its old role and start manufacturing red cells again.

4. *Destruction of hormones and drugs.* Many drugs and hormones are inactivated in the liver. Some are broken down completely while others are combined with another substance (such as glucuronic acid) which inactivates them. It is therefore important to be particularly careful when giving drugs to patients with liver disease. A normal dose for a normal person may be an overdose for such a patient because he cannot destroy the drug properly.

5. *Manufacture of clotting factors.* The liver is the place where most of the clotting factors essential for blood coagulation are manufactured.

6. *Manufacture of plasma proteins.* Many plasma proteins and in particular the plasma albumin are made in the liver.

JAUNDICE
Red cells are destroyed by the reticuloendothelial cells of the body, particularly in the liver, spleen and bone marrow. The haem part of the haemoglobin is converted into bilirubin which itself is insoluble in water. The bilirubin which is not made in the liver itself is therefore carried around to the liver in combination with the plasma proteins. Bilirubin in combination with plasma protein cannot be excreted by the kidneys and so none appears in the urine. This type of bilirubin is sometimes known as 'indirect' bilirubin because of its behaviour in the van den Bergh test for bilirubin in blood.

In the liver the bilirubin is extracted from the blood and combined (conjugated) with glucuronic acid to give bilirubin glucuronide. Bilirubin glucuronide is often known as conjugated or 'direct' bilirubin. Conjugated bilirubin is soluble in water and is excreted in the bile. Normally it does not enter the blood and so does not reach the urine. If it did gain entry to the blood, conjugated bilirubin could be excreted in the urine as it is freely soluble in water.

In the intestine, conjugated bilirubin is converted to a mixture of substances collectively known as urobilinogen. These are soluble in water and may be absorbed into the portal vein. However, the urobilinogen is normally promptly removed from the portal blood by the liver and re-excreted into the bile. Normally urobilinogen does not enter into the general circulation nor, even though it is soluble in water, does it enter the urine.

Jaundice is the term given to the yellow pigmentation of the skin, white parts of the eyes and mucous membranes, which occurs when the plasma contains unusually large amounts of either unconjugated or conjugated bilirubin. There are three main types of jaundice:

1. *Haemolytic*
This occurs when the red cells are being destroyed at an unusually rapid rate. Liver function is normal and the bili-

rubin in the blood is in the normal unconjugated form and is bound to plasma protein. No bilirubin therefore appears in the urine. No urobilinogen appears in the main circulation or in the urine, except occasionally when the rate of red cell destruction is so high that the liver cannot remove all the urobilinogen it receives in the portal blood. The faeces are usually very dark because of the excess bile pigment.

2. *Hepatogenous*
This occurs when the liver cells are damaged. There are many causes, the commonest being viral hepatitis, cancer and cirrhosis (destruction of the liver, sometimes due to excessive alcohol intake, but often of unknown cause). The flow of substances from the liver cells into the bile is disrupted and so some conjugated bilirubin, normally strictly confined to the bile, manages to enter the blood. Since conjugated bilirubin is soluble in water, it also appears in the urine. Furthermore, urobilinogen cannot be completely extracted from the portal-vein blood by the damaged liver and so it too appears in the general circulation and in the urine. This may not be true if the liver damage is very severe: then very little bile pigment reaches the gut at all and virtually no urobilinogen is formed so none can appear in the urine. Because of the lack of bile pigment the faeces are pale.

3. *Obstructive*
This is the result of mechanical obstruction of some part of the biliary tract, most often because of a gall stone but sometimes due to a tumour (e.g. of the pancreas). Some of the dammed up bile forces its way into the blood and so conjugated bilirubin again appears in the blood and the urine. There is not usually any urobilinogen in the urine: none is formed as no bile pigments can enter the gut because of the blockage. The faeces are usually very pale ('clay-coloured').

THE VITAMINS
During the nineteenth century it became clear that diets containing more than adequate amounts of highly purified fats, proteins and carbohydrates could not maintain health. Animals fed such diets developed a whole variety of diseases. Yet these diseases could be cured by the addition of minute quantities of substances known as vitamins to the food. Vitamins are organic compounds which are not oxidised by the body to supply energy, nor are they part of the structure of the body. They cannot be manufactured by the body in sufficient quantities and so they must be supplied in the diet. They are essential for normal metabolism and many of them, particularly those of the B group, act as coenzymes. This means that they work closely with an enzyme: they may bring to it atoms or small groups of atoms required for addition to a substrate or they may carry away similar atoms or small groups which may be split off from the substrate by the enzyme.

Many of the reactions for which vitamins are required are common to most of the cells in the body. An estab-lished cell which already has its full quota of vitamins will not initially suffer unduly if the supply of the vitamins to the body is reduced. Only if the deficient intake of the vitamin continues for a long time will established cells show any effects as the vitamin molecules within them are gradually destroyed in the process of wear and tear. In contrast to the established cells when the intake of a vitamin falls, those cells which are being newly manufactured will have inadequate amounts and will soon show the effects of the deficiency. Some tissues of the body are always manufacturing new cells to replace old ones which have been destroyed. The surfaces of the skin and the lining of the gut are quite quickly worn away and the cells there must be continually replaced by the formation of new cells. Similarly the red and many of the white cells of the blood are also continually being destroyed and must continually be replaced. Because of this most vitamin deficiencies first show themselves in the damage they do to the skin, the lining of the gut and the bone marrow where the red cells are manufactured. Virtually all vitamin deficiencies are therefore characterised by anaemia and by skin diseases.

When vitamins were first discovered, their chemical composition was unknown and so they were given letters, A, B, C, D and K. We now know the chemical composition of most of the vitamins and the letters are beginning to be replaced by the proper chemical names. However, it will be many years before the much simpler letters fall out of use.

The vitamins are usually divided into two great groups, those which readily dissolve in fats (A, D, E and K) and those which readily dissolve in water (B and C). The water-soluble ones are usually common in meat, cereals and vegetables while the fat-soluble ones occur primarily in dairy produce and fish-liver oils. The fat-soluble ones are absorbed from the gut in combination with fats and so anything which leads to a failure of fat absorption (e.g. lack of bile or pancreatic juice) leads to a deficiency of vitamins A, D, E and K.

THE B COMPLEX
The substance originally given the name of vitamin B is now known to be a mixture of many different chemically distinct compounds. However, these compounds often occur together in foods and when one is deficient the others are usually deficient too. Disease due to a pure deficiency of a single B vitamin is rare. The B-complex vitamins will therefore be briefly described, separately at first, and then the diseases which occur in B-group deficiency will be discussed at the end of the section. The common sources of each vitamin are listed in table 3.4.

Thiamine (Aneurin, vitamin B_1)
This is essential for many biochemical reactions but in particular for the conversion of pyruvic acid to acetyl coenzyme A. In thiamine deficiency, therefore, it is difficult to oxidise carbohydrate properly. Since the nervous system depends entirely on carbohydrate for its

Table 3.4 The main sources of vitamins in the food.

A	Dairy produce, fish-liver oils.
D	Dairy produce, fish-liver oils.
K	Widely distributed in small amounts. Made by gut bacteria.
E	Widely distributed, particularly vegetable oils and green vegetables.
Thiamine	Wholemeal flour, cereals, peas, beans, yeast.
Nicotinamide	Liver, meat, wholemeal flour.
Riboflavin	Meat, milk, wholemeal flour.
Pyridoxine	Very widespread.
Pantothenic acid	Liver, eggs, meat, milk.
Biotin	Widespread.
Folic acid	Widespread in small amounts. Made by gut bacteria.
B_{12}	Liver and all foods of animal origin.
C	Fresh fruits and vegetables.

energy supply, it is not surprising that it ceases to function properly in thiamine deficiency. The heart also depends heavily on thiamine and so there are defects in cardiac function. Thiamine is not destroyed by ordinary cooking but pressure cooking and canning remove much of it from the food.

Nicotinamide (niacin)

This too is essential for carbohydrate oxidation and for the operation of the citric acid cycle. It is very resistant to cooking. It is only a partial vitamin in the sense that the body can manufacture part of its daily requirement from the amino acid tryptophan. Only if the diet is deficient both in nicotinamide and tryptophan will signs of deficiency appear.

Riboflavin

This also is required for oxidation of foodstuffs. With most diets an important part of the supply comes from milk. It is moderately resistant to cooking but is rapidly destroyed by exposure to bright light. Ninety per cent of the riboflavin in a bottle of milk may be destroyed if the bottle stands in the sun for a morning. In man another important source of riboflavin is the bacterial colony in the intestine which manufactures it. Sterilisation of the gut by antibiotics may greatly reduce the riboflavin intake.

Pyridoxine (vitamin B_6)

This is important for the manufacture and interconversion of amino acids and also for the utilisation of the essential fatty acids. It is very widespread in ordinary foods but some tropical diets and some synthetic infant diets may be deficient in it. Isoniazid, a drug important in the treatment of tuberculosis, can inactivate pyridoxine. Signs of vitamin deficiency may therefore appear even though the intake in the food is normal.

Pantothenic Acid, Lipoic Acid and Biotin

Most diets contain adequate amounts of these vitamins and so deficiencies are highly unlikely. Biotin cannot be absorbed from the gut in the presence of raw egg white.

The only recorded case of deficiency occurred in a man whose diet consisted entirely of a dozen raw eggs and several pints of wine daily!

Folic Acid

This is essential for the manufacture of nucleic acids and so is particularly required when cell division and protein synthesis are occurring rapidly. It is widely distributed in the food but much is made in the gut itself by the resident bacterial colony. It is particularly required by the skin and the lining of the gut, by the marrow manufacturing red cells and by the pregnant mother acting as host for the growth of a new child. Deficiencies are most likely to occur in infancy, before the normal bacterial colony has been established, and in pregnancy when requirements are very high. A deficiency results in a type of anaemia in which not enough red cells can be manufactured. Haemoglobin synthesis is less affected and so there is too much haemoglobin for the number of cells being formed: the cells therefore tend to be larger than normal. This type of anaemia is known as megaloblastic. It is also not uncommon in epileptics: many of the drugs used in epilepsy interfere with action of folic acid. Oral contraceptives may also interfere with it. In the tropics there may be changes in the gut bacteria which leads to the colony using more folic acid and manufacturing less. The deficiency damages the lining of the gut leading to a condition known as tropical sprue, one of whose symptoms is persistent diarrhoea.

Rapidly dividing cancer cells also need large amounts of folic acid. Drugs which interfere with the action of folic acid are therefore sometimes used as anti-cancer agents.

Vitamin B_{12} (cyanocobalamin)

This is an unusual vitamin in that it is very widely distributed and is required in such minute amounts that natural diets are never deficient in it. The only people likely to suffer from a dietary deficiency are those extreme vegetarians known as vegans who do not eat any animal products (including milk and eggs). In spite of this, disease due to vitamin B_{12} deficiency is very common. This paradoxical situation arises because a special absorptive mechanism is required to get the vitamin across the gut wall into the blood. The stomach

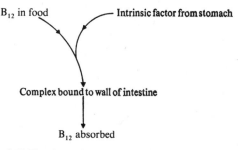

Figure 3.19 The absorption of vitamin B_{12} from the gut.

secretes a substance known as intrinsic factor which combines with the vitamin B_{12} in the food. The combination of intrinsic factor and vitamin is then bound to the wall of the small intestine and the vitamin is taken across into the blood. In the absence of intrinsic factor the vitamin cannot become bound to the wall of the intestine and so is hardly absorbed at all even though there may be plenty in the diet. A failure of intrinsic factor secretion is not uncommon and gives rise to the disease known as pernicious or Addisonian anaemia.

Vitamin B_{12} like folic acid, is required for cell division and so the anaemia is of the megaloblastic type. Megaloblastic anaemia may therefore be caused by a deficiency either of folic acid or of B_{12}. Pernicious anaemia is caused only by B_{12} lack: in addition to the anaemia, peripheral nerves and parts of the spinal cord degenerate. Vitamin B_{12}, unlike folic acid, seems to be essential for the functioning of these parts of the nervous system.

B-Complex Deficiencies

There are two major diseases which are caused by B-complex deficiency. Beri beri is primarily due to thiamine deficiency but is always aggravated by lack of other B vitamins. Pellagra occurs primarily where corn (maize) is the staple diet and is thought to be due to a combined absence of nicotinamide and tryptophan although again the lack of other B-complex members almost certainly plays a part.

Beri beri itself occurs in two main forms. In the 'dry' variety, the main features are due to defects in the nervous system. Muscles become wasted and paralysed and sensation is lost. The brain may also be damaged, resulting in lethargy, loss of appetite and memory loss. In developed countries, beri beri is likely to occur only in chronic alcoholics who gain all their calories from alcohol and who eat no proper food. In the 'wet' form of beri beri, the main organ to suffer is the heart. This cannot pump blood around the body effectively. As a result the kidneys fail to excrete fluid normally and large amounts of water accumulate in the form of oedema.

With pellagra, the early signs of the disease are seen in the skin and in the lining of the gut. The mouth and tongue become red and ulcerated and there is vomiting and profuse diarrhoea. Muscles become weak and sensation is affected. Finally the brain is damaged and the patient becomes frankly mad. As with beri beri the end result of the untreated disease is death.

VITAMIN C (ASCORBIC ACID)

Scurvy is a disease which has been well known for hundreds of years. It has always tended to occur in people deprived of fresh food for long periods such as those going for long sea voyages and those in northern cities in winter. Today in the developed countries the disease may sometimes be seen in lonely old people who eat everything out of tins. Scurvy is due to a deficiency of vitamin C. The vitamin is found primarily in fresh fruit and vegetables. Since it is very easily destroyed by cooking or canning, cooked food may contain very little

of it. Freezing, in contrast, has little effect on the vitamin C content of food.

The main effects of ascorbic acid deficiency can be accounted for by the fact that it is essential for the manufacture of the structural protein, collagen. Collagen is important in bones and joints, in tendons and ligaments, in the support of blood vessels and of the skin and in the healing of wounds (white scar tissue consists largely of collagen). It is therefore not surprising that the main features of scurvy are bleeding around the gums and into the skin, painful joints and a failure of wounds to heal quickly. In children the skeletal system fails to develop normally because of the lack of collagen for bone. In addition to its role in collagen synthesis, vitamin C appears to be essential for the normal function of folic acid. Thus anaemia is also a common feature of scurvy.

VITAMIN A

Vitamin A is found in large amounts in dairy produce and in fish-liver oils. In addition, all green or yellow plants and vegetables contain substances called carotenes which are not active vitamins themselves but which can be converted to vitamin A in the body. The vitamin is stored in large amounts in the liver. The main consequences of vitamin deficiency are as follows:

1. Night blindness develops. Vitamin A is essential for the normal functioning of the rods, the elements of the retina at the back of the eye which are required for vision in dim light. The vitamin is not required for the functioning of the cones with which we see when the light is bright. The earliest complaint of the patient is often therefore inability to see in the dark.

2. The vitamin is essential for the normal moulding of bone in growth. In the absence the skull, and the holes in it which allow the nerves to pass out and in, fail to enlarge normally. The brain and the nerves therefore become damaged.

3. The skin and the mucous membranes lining most of the tubes in the body (e.g. the gut and the bronchi of the lungs) become thick, dry and liable to infection. The cornea of the eye becomes cornified and blindness may occur. Digestion is poor and because of damage to the respiratory tract, pneumonia is a common cause of death.

4. Males are infertile and while females can conceive they cannot carry normal infants to term.

In these days of readily available enriched sources of vitamins, it is not impossible, especially in infants, for an overdose to occur. The characteristics are an itchy rash, severe headache, weakness, loss of appetite and pain in the long bones. The condition is reversed within a few days of stopping the high intake of the vitamin.

VITAMIN D

Vitamin D_2 (calciferol) and vitamin D_3 are closely related compounds, both of which have vitamin activity. There is also another substance, 7-dehydrocholesterol, which is common in dairy foods and fish-liver oils. In itself it is inactive but it is converted to vitamin D_3 on exposure to ultraviolet light: the conversion can readily occur in skin exposed to sunlight.

Vitamin D is important to the body in two main ways:

1. It is essential for the absorption of calcium from the gut. In its absence sufficient calcium cannot be absorbed to replace even the small calcium losses which occur into the urine and faeces in adults. In pregnancy, and in infancy and childhood when calcium requirements are high for lactation and new bone formation, the lack of calcium is particularly important.

2. It is essential for the manufacture of normal new bone and for the incorporation of calcium into bone.

Two main types of disease can result from a lack of vitamin D, rickets in children and osteomalacia in adults. Rickets is a disease in which the growing points (epiphyses) of bones do not function properly and as a result the bones become twisted. This leads to permanent deformities which persist into adult life. In adults whose epiphyses are no longer active, rickets cannot occur. However, the persistent lack of calcium leads to softening of the bones and particularly of the pelvis which has to bear so much of the weight of the body. The deformed pelvis may narrow the outlet for the fetus and cause difficulties in childbirth.

When rickets was first being studied, there was considerable argument about its cause. One group maintained that lack of sunlight in narrow, smoke-blanketed city alleys was the cause. The other said that rickets was due to a poor diet. In fact it is now apparent that both ideas were correct. If there is enough D_2 and D_3 in the diet, rickets will not occur. On the other hand, 7-dehydrocholesterol is found in the skin: if the skin is exposed to ample sunlight, again enough D will be formed and rickets will not occur.

Vitamin D is perhaps the vitamin which is most likely to be deficient in what seems to be a normal diet. This is especially true in infancy and in pregnant and lactating women. Margarine is therefore artificially enriched with the vitamin in most countries and it is usually a routine precaution to give extra amounts of the vitamin to infants and to nursing and pregnant mothers.

VITAMIN K

Vitamin K is a blanket expression which covers a number of substances. Most of these are naturally occurring but one, K_3 or menadione, is artificially synthesised. Vitamin K is widely distributed in food and quite large amounts are manufactured by the bacteria in the gut. The only situations when K deficiency is likely are:

1. In the first days of life when there are no bacteria in the gut. Infants in whom there is any risk of bleeding (e.g. due to birth trauma) are routinely given an injection of vitamin K.

2. Following sterilisation of the bowel by antibiotics.

3. In obstructive jaundice or in any other condition when the absorption of fatty substances as a whole is defective.

4. In the presence of drugs which interfere with the action of vitamin K. These occur naturally in some plants and cows have become K deficient after eating these. Drugs based on the substances isolated from these plants are now deliberately used in medicine to interfere with blood coagulation in situations in which there is an abnormal tendency for clotting to occur.

The main action of the vitamin appears to be in the synthesis of clotting factors by the liver (particularly prothrombin and factors VIII, IX and X). In the absence of these factors, clotting is defective and abnormal bleeding is likely to occur.

VITAMIN E

This too is a blanket expression for a group of compounds known as the tocopherols. They are very widely distributed and a natural deficiency is most unlikely except possibly in those who over a very long period are incapable of absorbing fat. The main effects of deficiency which have been studied in animals and man are infertility in both sexes, muscle weakness and anaemia. There is no evidence that human infertility is ever due to vitamin E deficiency.

NUTRITION

In designing a good diet, at least four major problems must be dealt with.

1. Sufficient water must be provided daily. No one ever thinks of this as a nutritional problem yet it should be remembered that a lack of water kills in days whereas a lack of food usually takes weeks to have this effect.

2. Foodstuffs which can be oxidised to provide energy must be supplied in adequate amounts. The energy requirement is usually expressed in terms of kilocalories. Most of the energy comes from the oxidation of fat and carbohydrate with much smaller amounts coming from protein.

3. A diet which supplies enough calories may still not contain all the substances required for growth in children and the maintenance of full health in adults. The essential amino acids, essential fatty acids, minerals and vitamins must also be present.

4. A diet which completely fulfils the above three conditions may nevertheless still be defective. Instead of containing too little of a substance it may contain too much.

Natural substances which may cause disease when taken in excess include fat, carbohydrate and vitamins A and D. But increasingly the food contains unnatural substances which should not be there. Some of these, such as the insecticides, get there purely by accident: they are so widely distributed on the earth's surface that it is impossible to stop them getting into the food. Others such as cyclamates and saccharin may be put in as sweeteners, preservatives or colouring matter. In most cases we do not know whether or not these substances are harmful.

The second and third problems face the less developed areas of the world. The fourth is primarily a problem met with in the developed countries.

CALORIFIC REQUIREMENT AND METABOLIC RATE

The numbers of calories released by pure samples of the various types of food can be measured by burning a known weight of the food in a device known as a bomb calorimeter. It has been demonstrated that when food is burned in such a device it releases precisely the same number of calories as does the same amount of food when oxidised in the body itself. Each gram of protein burnt yields about 5·3 kcal. Fat yields about 9·3 kcal/g while carbohydrate yields about 5·3 kcal/g. Thus a given weight of fat releases more calories than a given weight of any other food and it is therefore not surprising that excess food is stored in the form of fat.

There are a number of ways of measuring the number of kilocalories worth of food which the body consumes. The simplest one depends on the fact that for ever litre of oxygen used up by the body, about 4·8 kcal are produced. The person under test is asked to breathe from a bag or a cylinder containing a known amount of oxygen. After a measured time, the amount of oxygen left is measured and so the amount of oxygen used up can be easily calculated. The number of kilocalories produced in that time can then be worked out by multiplying the number of litres of oxygen used by 4·8.

The metabolic rate of a person is the number of calories he or she uses up per day when carrying out normal activity. The number of calories taken in in the food can be estimated by a careful analysis of the diet: the weight of fat in grams is measured and multiplied by 9·3 to give the number of kilocalories which could be produced by the oxidation of that amount of fat: the number of kilocalories which could be released by carbohydrate is estimated by measuring the weight of carbohydrate in grams and multiplying it by 5·3: the potential energy yield of the protein is estimated in the same way. Thus the amount of calories which could be released by the amount of food taken in can be compared with the number of calories actually used by the body each day. We can represent the balance between the two by means of the following equation:

Calorific value = Number of calories + Calorific value
of food   used by the body   of fat stored

This means that if the food contains more calories than are used up in a day by bodily activity, the extra calories are stored away as fat. If the food contains too few calories to supply the energy needs for that day then some of the fat stores are broken down to make up the difference. It is impossible to avoid the consequences of this equation. There is no easy road to the loss of weight. The only way to lose weight is to reduce the food intake below the level of energy expenditure. This can be done either by taking the same amount of exercise but eating less or by eating the same amount but carrying out more exercise, or of course by combining the two things. It is as simple (or as difficult) as that. Those who are fat eat too much in relation to the amount of exercise they do. Those who are put on a diet and still do not lose weight are either still eating too much or not exercising enough. Weight will always be lost if calorie expenditure rises above calorie intake. No tricks can be used to get around this fundamental relationship.

FACTORS WHICH ALTER METABOLIC RATE

As mentioned earlier, the metabolic rate of a person is the number of calories used up per day. The basal metabolic rate is something rather different which attempts to express something more fundamental about the metabolism of the body. For example, the crude metabolic rate of a large healthy man might be in the region of 3000 kcal/day while that of a healthy child might be 1000 kcal/day. That simply means that more calories are required to run a large body and does not imply that there is something fundamentally different between the metabolism of a man and that of a child.

In trying to estimate the basal metabolic rate (BMR), the first thing is to eliminate differences which occur because of differences in muscular activity and food intake, by keeping the person at rest and without food for at least 12 hours. This will bring the calorie expenditure of the lumberjack and the businessman down to the same level but still they will have much larger crude metabolic rates than a child. It has been found, however, that the BMR is closely related to the surface area of the body. The number of kilocalories used up in a resting individual per square metre of body surface is remarkably constant for all healthy humans and indeed for all healthy mammals. The BMR is therefore expressed as the number of kilocalories produced per square metre of body surface per day. The normal figure is close to 1000 $kcal/m_2/day$. It can easily be measured indirectly by measuring oxygen consumption.

The following factors may alter the BMR:

1. It falls very slowly with increasing age.

2. It is very slightly lower in women than in men.

3. It may increase on prolonged exposure to cold weather. The body thus produces more heat for the maintenance of its normal temperature.

4. It rises by about 10% for every 1 °C rise in the body temperature. The BMR therefore may be greatly elevated in fever.

5. It is affected by the level of thyroid hormone. When the output of thyroid hormone is too high (thyrotoxicosis) the BMR may be 50% or more above normal. When the hormone output is too low (myxoedema), the BMR may be well below normal. BMR measurements used to be made during the diagnosis of thyroid disease but they have now largely been superseded by other ways of measuring thyroid activity.

In normal individuals, most of the above factors which can alter metabolic rate are overshadowed by the much larger changes caused by food intake and exercise which by definition are excluded from the BMR. Even if an individual remains completely at rest, taking in food increases his metabolic rate. This is known as the specific dynamic action of food and it seems to occur because energy is required for the digestion, absorption and processing of food. The energy used up in this way, in making the food available for use by the body, is clearly not available for other bodily activities. Thus only about 90% of the calorific value of food can actually be used by the body. Even the rise in metabolic rate on taking in food is totally overshadowed by the extra calories required for even mild exercise. Work at an office desk requires an extra 40 kcal/hour above basal levels while the shovelling of earth might use up an extra 300 kcal/hour or even more. Therefore in calculating the total calorific requirement of an individual the following factors must be taken into account:

1. BMR required to keep the body ticking over. For an average sized man this might be 1700 kcal/day.

2. Extra calories required because of specific dynamic action of food. An extra 10–15% over the basal level would be required, say an extra 200 kcal, making 1900 kcal/day.

3. Extra calories required for muscular activity. Assume that the person spends 8 hours sleeping, 8 hours in leisure activity and 8 hours working.
   a. Sleep requires no extra calories.
   b. Leisure energy expenditure clearly varies widely, but a reasonable estimate might be an extra 50 kcal/hour. Eight hours leisure would therefore add 400 kcal making 2300 in all.
   c. Eight hours work. An office job might add 8×40 = 320 kcal, while lumberjacking might add 8×300 = 2400 kcal. This therefore produces the really big differences in calorific requirements which range from about 2600 kcal/day for a light office job, to about 4700 kcal/day for heavy labour.

4. Finally 10–15% of the calorific value of unprepared food is lost, being destroyed by cooking, left on the side of the plate or passing straight through the gut to the faeces. Therefore the calorific value of the unprepared food must be 10–15% higher than the calorific requirement of the body.

SUPPLYING CALORIES
Approximate daily calorific requirements of children and of adults doing various sorts of work are shown in table 3.5. In developed countries about 10–15% of the calories

Table 3.5 Some approximate daily kilocalorie requirements for children and adults.

Children, 1–2 years	1,000
Children, 2–3 years	1,250
Children, 3–6 years	1,550
Children, 6–8 years	1,850
Children, 8–10 years	2,150
Children, 10–12 years	2,550
Children, 12–14 years	2,900
Girls, 14–18 years	2,900
Boys, 14–18 years	3,200
Housewives	2,700
Nurses, men doing light work	3,000
Men doing heavy work	4,000

are supplied by protein. The remainder are supplied by fat and carbohydrate, the relative proportions of the two depending to some extent on the economic status of the individual. Carbohydrate-rich foods tend to be much cheaper than fats. Many poorer people therefore obtain almost all their calories from such foods as bread and potatoes. With increasing income, fat-rich foods such as dairy products and meat, can be bought in larger amounts and as much as 40–50% of the total calorie requirement may be supplied in the form of fat. In most developed countries, the problem is that most people take in too many calories for the amount of exercise that they do and hence tend to become obese. In less fortunate areas of the world, a low calorie intake is much more significant.

STARVATION
Most of the people in the poorer parts of the world are malnourished rather than completely starving. They receive neither enough calories nor enough of essential dietary constituents such as amino acids, vitamins and minerals. A number of experiments have revealed the devastating effects in adults of a simple reduction in calorie intake to the level of 1500 kcal/day. The human volunteers in these experiments were previously completely healthy and their diet contained more than enough protein, vitamins and minerals. The only defect was a lack of calories. Quite soon these previously fit men and women became weak, lost weight, and were incapable of making any sustained mental or physical effort. It is frightening to think of the consequences when a whole nation is in this state of health.

Although undernourishment is much commoner than total starvation, the latter is by no means unknown. If the intake of water is normally maintained, a normal lean individual will die within 4–6 weeks if he is totally cut off from food. The stages in the process are as follows:

1. Glycogen stores are used up within the first couple of days.

2. The body must then rely on fat oxidation for its energy supply. There may not be enough oxaloacetic acid for the entry of acetyl coenzyme A into the citric acid cycle. The ketone bodies then tend to accumulate and ketosis may occur.

3. Protein is broken down only very slowly but its degradation does occur and the muscles which are largely protein gradually lose their bulk. When the fat deposits have been finally used up, protein is the only food left. It must therefore be broken down much more rapidly and there is a sharp increase in the rate of production and excretion of urea. This is known as the 'premortal rise' because it indicates that death cannot be far off.

One of the outstanding features of starvation is the accumulation of oedema fluid in the legs and abdomen. The bloated abdomen looks particularly tragic against the background of the remainder of the emaciated body. The cause of the oedema is uncertain. Part may be due to the weakness of the heart which prevents the normal excretion of fluid by the kidneys. Part is due to the fall in the plasma protein concentration which occurs during starvation. The osmotic pressure of the blood thus falls and this allows fluid to escape into the tissues.

KWASHIORKOR
No human disease is known to occur because of a deficiency of a single amino acid. However, a generally low protein intake is of major importance in the disease kwashiorkor which is not uncommon in South America, Africa and Asia. The work is said to mean 'displaced child' for the disease usually develops when a child is prevented from breast feeding by the arrival of a younger infant. The outstanding characteristics of the disease are retardation of growth, anaemia, fatty degeneration of the liver and a deficiency of pancreatic enzymes. In Negro children the hair is usually reddish instead of its normal black colour. Most of the changes are obviously related to protein deficiency. The lack of digestive juices obviously clearly makes things worse for it is impossible to digest and absorb properly even the small amounts of protein which may be taken in with the food. In adults protein deficiency leads to muscle wasting, liver degeneration, oedema and a hopeless feeling of lassitude.

ESSENTIAL DIETARY CONSTITUENTS
Apart from the necessity of providing enough calories, all diets must contain certain additional materials which the body cannot manufacture for itself. The main ones will be briefly discussed in this section.

Essential Amino Acids
There are ten amino acids which must be provided in the diet because they cannot be manufactured by the body in sufficient quantities to meet its needs. Proteins which contain all the essential amino acids are said to be of high biological value. Those which do not contain them all are said to be of low biological value. In general, proteins of animal origin such as those of meat and milk are of high biological value. Those of vegetable origin tend to be of low biological value. However, two different vegetable proteins are not usually lacking in the same amino acids and they can therefore make up one another's deficiencies. Thus, two vegetable proteins of low biological value may together make up a meal of high biological value. It is important to realise that all the essential amino acids must be present in the same meal if they are to be used for protein synthesis. Amino acids which cannot be used at once are promptly broken down by the liver and wasted. Therefore it is pointless to give a protein of low biological value at one time and then another different low-biological value protein 3 hours later. The amino acids from the first one will be broken down before the second arrives on the scene.

Elements
The main elements which must be provided in the food are sodium, potassium, chlorine, calcium, phosphorus, iron, iodine, magnesium, copper, manganese and fluorine. Of these, calcium, iron, iodine, magnesium and fluorine are particularly important because deficiencies can occur in practice as well as in theory. Deficiencies of the other substances are extremely unlikely to occur naturally. The main situations in which deficiencies of elements may occur are as follows:

1. Iron is essential for the manufacture of the haemoglobin in the red cells of the blood. It is also required in much smaller amounts for all other cells. As cells are repeatedly being lost from the surfaces of the skin and gut, about 1 mg of iron/day is lost from the body in this way. The average western type diet contains 15–20 mg of iron/day but absorption from the gut is poor and only about 1.5–2 mg of iron enter the body. The absorption of even this small amount seems to depend on the presence of normal amounts of acid in the stomach. People with chronic inflammation of the stomach (chronic gastritis), those with gastric ulcers, and those who have had part of their stomach surgically removed produce less acid than normal. They cannot therefore absorb iron normally and are likely to develop an iron deficiency anaemia unless extra iron is provided in the diet. Because of loss of iron in the blood during menstruation, all women of reproductive age are also likely to develop a similar anaemia. With most women the amount of blood loss when averaged out over the whole month means an extra loss of about 1 mg of iron/day. Women therefore lose about 2 mg of iron/day in all and are in a precarious state of iron balance.

2. Calcium is primarily found in dairy products and most adults in the western countries probably get sufficient. However, children and pregnant and lactating women need more calcium and must take in extra

supplies, usually in the form of milk. It has recently become apparent that old people living alone also tend to take in too little milk and may become calcium deficient. Their bones may weaken more quickly than is usual in old age.

3. Iodine is essential for the manufacture of thyroid hormone. If there is too little iodine in the diet, the person may become thyroid deficient and the gland may grow in a desperate effort to produce more thyroid hormone. This produces the unsightly swelling in the neck known as a goitre. Iodine occurs on this planet primarily in the sea, and on moving away from the sea, drinking water and food contain less and less iodine. Goitre therefore tends to occur in mountainous areas far from the sea such as Switzerland, the Himalayas and Derbyshire in England. In many countries, iodine is added to salt to ensure that no one suffers from a goitre of this type.

4. The importance of magnesium has been recognised only recently. It seems to be essential for the normal functioning of the nervous and muscular systems and in its absence there may be spasms, irritability and eventually unconsciousness. Most normal diets contain more than enough magnesium and deficiency is likely only in two abnormal situations. The first is severe diarrhoea when quite large amounts of magnesium may be lost in the faeces. The second is during the prolonged maintenance of a patient on intravenous fluids which do not usually contain magnesium salt.

5. Fluoride seems to be essential in only one way, for the development of strong teeth which are resistant to decay. Most natural water supplies contain less than the optimum amount of fluoride to achieve this end and so in many parts of the world small amounts of fluoride are added to the drinking water in order to enable the body to resist tooth decay more effectively.

Essential Fatty Acids and Vitamins
These have already been discussed in other sections.

PRACTICAL NUTRITION
In planning a diet for a patient, four conditions must be fulfilled:

1. The diet must contain enough calories. Normal daily calorie requirements are shown in table 3.5 above and the calories contents of some well-known types of food in table 3.6.

2. The diet must contain enough protein of high biological value, enough vitamins, enough elements and enough essential fatty acids.

3. It must contain above-normal amounts of some substances in special cases, e.g. pregnant women may need extra iron, calcium and folic acid.

Table 3.6 The kilocalorie contents of 100 g of various types of food.

White bread	240
Rice	360
Milk	70
Butter	790
Cheese	420
Steak	270
Fish	170
Potatoes	70
Peas	80
Cabbage	10
Orange	30
Apple	50
Sugar	400

4. Some patients may be harmed by certain foods and in these cases a diet which does not contain these foods must be provided.

Some special nutritional problems which occur frequently are discussed in the remainder of this chapter.

Obesity
Any observant nurse or medical student very quickly learns the dangers of being overweight. Many of the patients in the wards suffer from obesity and this may aggravate or even be the cause of their other diseases. The insurance companies have demonstrated conclusively that the fatter a patient is, the more likely is he or she to die at an early age. Obesity plays a contributory role in many conditions, notably heart disease and high blood pressure. No one who has been in an operating theatre needs to be reminded of how much more difficult is the surgeon's task when the patient is obese. Postoperative respiratory problems are also much more common in fat people who find it more difficult to breathe deeply and evenly. Arthritis tends to be commoner in the fat as the joints, especially the legs, have to carry so much more weight. There are virtually no advantages in being fat.

Any obese person has been eating too much for the amount of exercise he or she has been doing and the extra calories have been laid down as fat. Very few obese patients have any hormonal abnormalities which cause their fatness and their state is the result either of gluttony or lack of exercise. The aim of the dietary treatment of obesity is therefore to reduce the calorie intake below the patient's calorie needs so that the extra calories can be obtained by burning up the fat deposits. A drastic reduction in calorie intake can make a person feel weak and ill. On the whole it is better for a patient to lose weight slowly but steadily. It is better to acquire a habit of eating more wisely than it is to attempt a crash diet which is unlikely to be of any permanent value. The crash diet may be briefly successful but it usually makes the patient feel ill. Moreover, since it is not continued long enough to bring about any permanent change in eating habits, the weight is usually rapidly regained as soon as the crash period is over. For all these reasons, working people who want to lose weight should probably attempt a 2000 or even a 2500 kcal diet. This will enable them to lose weight slowly but consistently without becoming ill; 1500 or 1000 calorie diets should be attempted only by those who are not working and can afford to rest.

Pregnancy

The weight gain in a normal pregnancy is not usually much more than about 60 g (roughly 2 oz)/day. It is therefore quite unnecessary to have a great increase in calorie intake. However, some particular essential food factors, notably iron, calcium and folic acid, tend to get used up much more rapidly and it is wise to take supplements of these. Taking extra vitamins is unlikely to do any harm although for most women on a good diet in developed countries there is little evidence that this is necessary.

Diabetes

The main problem in diabetes is that the body cannot cope normally with carbohydrates. The aim of a diabetic diet is therefore to reduce the intake of carbohydrate. Since many diabetics, especially elderly ones, tend to be obese, the diet usually also aims to bring down body weight. The most dangerous carbohydrates are those like sucrose (cane sugar) which need very little digestion or those like glucose which do not need to be digested at all. These substances are very rapidly absorbed from the gut and the sudden flood puts a great strain on the diabetic's limited ability to cope with carbohydrate. An attempt is therefore usually made to exclude cane sugar from the diet as completely as possible. The only carbohydrates permitted (and these usually in small amounts) are ones like starch which are slowly digested and which enter the blood in a steady trickle.

Renal Failure

When the kidneys fail to excrete waste products normally, the body has great difficulty in getting rid of potassium and of urea and other products of protein breakdown. The protein breakdown products affect the nervous system and cause weakness and a comatose state: this is often called uraemia because the blood concentration of urea becomes very high. The potassium affects the heart and may interfere with its normal rhythm.

The major sources of potassium in the diet are fruit and meat and so the intake of these must be cut down. The protein intake must be cut down to the minimum compatible with the maintenance of life. This reduces the rate of urea accumulation. The protein must of course be provided in a form that is of high biological value and the patient with renal failure presents a real challenge to the dietician.

Effects of Drugs

A few drugs must be considered in diet. The most important are the antidepressive agents, the monoamine oxidase (MAO) inhibitors. Monoamine oxidase is an enzyme which normally seems to destroy naturally occurring amines in the body (such as adrenaline and noradrenaline) and also some of the amines which occur in food. Some of these food amines act to release large amounts of adrenaline and noradrenaline. In the absence of the enzymes to destroy the amines, a substance such as tyramine which occurs in a number of foods can produce a dramatic attack of high blood pressure, with palpitations and a throbbing headache. Patients on these drugs must therefore be carefully instructed to avoid certain foods, including cheese, bananas and hydrolysed meat extracts.

Food Allergies

Some unfortunate people are highly sensitive to substances such as eggs or sea foods. If they accidentally eat the forbidden material they may feel weak and ill and come out in an unpleasant rash. The cause of this situation is by no means fully understood. However, it is known that the body can recognise proteins which are foreign to it. It can mount what is called an immune response against the foreign protein and destroy it and this behaviour is of great value in the resistance to bacterial infections. Unfortunately, as a side effect of the immune response, many tissues in the body may become damaged and the patient may become ill.

Normally, of course, protein is completely broken down in the gut. Protein as such cannot be absorbed and gain access to the body: it is first broken down to amino acids which cannot provoke an immune response. Some unfortunate individuals seem to have unusually permeable gut walls and so some proteins may be able to pass through without first being broken down. These proteins can then provoke an immune response. The patient therefore develops an allergy to such proteins and must strictly avoid eating any food which contains them.

Metabolic errors

Some patients have genetically transmitted inborn errors of metabolism. That is to say they lack certain enzymes essential for particular biochemical reactions. A number of these errors present dietary problems. The main ones of importance are:

1. *Coeliac Disease.* Wheat flour contains a protein known as gluten. In some individuals this cannot be fully digested and one of the products of its partial digestion is highly toxic to the wall of the intestine. The gut wall is so damaged that it ceases to be able to absorb a whole variety of materials. Much of the food passes straight through, causing diarrhoea and malnutrition. The disease can be cured by a diet which contains no wheat or wheat products and the gut wall then quickly returns to normal.

2. *Disaccharidase deficiency.* There are a number of enzymes in the gut which are essential for the breaking down to monosaccharides of the disaccharides sucrose, lactose and maltose. A deficiency of any one of these enzymes can lead to diarrhoea and a failure of absorption. If it is lactose which cannot be digested, the condition appears almost immediately after birth when the infant takes milk. It can be cured by giving a lactose-free

diet, which of course means avoiding milk. Sucrose enzyme deficiencies tend to become apparent later when sucrose is first given: they can be cured by eliminating sucrose form the diet. In these cases, enzyme systems often develop as the child grows older and so he grows out of his disease.

3. *Galactosaemia.* Some infants cannot convert galactose to glucose. This is very important because half the carbohydrate taken in is in the form of galactose from the lactose of milk. The galactose accumulates and can cause severe brain damage unless it is excluded from the diet.

4. *Phenylketonuria.* The body is unable to deal with phenylalanine normally. The amino acid accumulates and abnormal substances derived from it damage the brain. The situation can be improved by giving a diet low in phenylalanine. This is very difficult to do and very expensive because phenylalanine is found in most proteins.

Intravenous Feeding

Many patients in hospital cannot digest and absorb food normally for a variety of reasons and so they must be maintained by dripping fluid directly into the blood. The important points to remember are:

1. Adequate amounts of fluid must be given. If the patient's kidneys are functioning normally it does not matter if rather too much water is given as the excess can easily be eliminated. But if the patient's kidneys are failing then it is vital to calculate the fluid requirement very carefully as it is easy to give too much water and overload the circulatory system.

2. Sufficient calories must be provided. This is not easy to do, especially if the patient is in renal failure, so restricting the amount of fluid which can be dripped into the body. Highly concentrated glucose solution can be employed, but this often clots the blood in the vein unless the drip is so arranged that it enters a major vessel like the superior vena cava where it can be rapidly diluted. In recent years liquid-fat drips have become available and these can provide a high calorie content in a small volume of liquid.

3. Protein can be provided in the form of predigested amino acids.

4. Elements. Particular care must be taken of the balance of intake and output of sodium, potassium, calcium and magnesium. It is all to easy to provide too little or too much of these elements.

5. Vitamins must be provided by intramuscular injection.

4 The Nervous System

The nervous system is ultimately responsible for the control of all the other organs of the body. As such it is obviously of the greatest importance. Its most important components are:

1. *Sensory receptors*
The function of these is to collect information about the state of the body itself and also about the surrounding environment. For example, the eye and the ear gather information about the environment by using light and sound, while receptors in the skin collect information about objects coming into contact with the body. Receptors within the body itself include ones for measuring blood pressure, body temperature, blood glucose concentration, the positions of joints, the states of contraction of muscle and many other things.

2. *Sensory nerves*
The information collected by the receptors which are spread throughout the body must be carried to the control centres in the brain and spinal cord. This is done by means of electrical changes known as nerve impulses which are carried from the receptors to the central nervous system along sensory nerves. Sensory nerves are sometimes called afferent nerves.

3. *Central nervous system (CNS)*
The central nervous system consists of the brain and the spinal cord. All the information from receptors is carried to the CNS where its significance is assessed. The CNS then makes a decision as to what action if any is required. The CNS can also store much of the information it receives in the form of memory.

4. *Effector nerves*
Effector nerves carry instructions from the CNS to the various organs in the body. Again these instructions are in the form of nerve impulses travelling along nerve fibres. Effector nerves are sometimes known as motor nerves or as efferent nerves.

5. *Effector organs*
These carry out the instructions of the nervous system. The most important effectors are:

a. Skeletal muscle. This is sometimes known as striated muscle. As its name suggests its main function is to move the skeleton.

b. Smooth muscle. Smooth muscle is the type found in the internal organs such as the gut, the urinary tract, the walls of blood vessels and the internal genital organs.

c. Cardiac muscle in the heart.

d. Exocrine glands which secrete fluids along ducts on to some body surface. Examples of exocrine glands are the pancreas and the salivary glands which secrete digestive juices, the lachrymal glands which secrete tears and the sweat glands which secrete sweat.

e. Endocrine (or ductless) glands which secrete chemicals known as hormones into the blood flowing through the gland. The hormones are then carried by the blood to all parts of the body and as will be seen later, they themselves control the behaviour of many effector organs. For example, the behaviour of important effector organs like the liver and kidney is to a large extent determined by hormones.

NERVE CELLS
The nervous system is made up of nerve cells or neurons as they are often called. Each neuron, like almost all other cells has a cell body containing a nucleus and cytoplasm. Neurons are however distinguished by having long fine processes arising from the cell body. Short ones which end close to the cell body are known as dendrites: they form the input end of the cell and nerve impulses arriving reach the cell by means of the dendrites. Long ones which travel some distance from the cell body are known as axons: they are the output side of the cell and nerve impulses generated by the cell leave by the axon.

There are four main types of nerve cell:

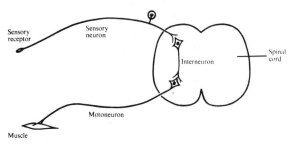

Figure 4.1 The main types of nerve cell.

1. CNS nerve cells

These consist of a cell body, dendrites and an axon. The axon carries impulses from one part of the CNS to another.

2. Somatic motoneurons

These carry instructions from the CNS to the skeletal muscles. They consist of a cell body inside the CNS which gives rise to a very long axon. This axon leaves the CNS and carries nerve impulses to the skeletal muscles.

3. Autonomic motoneurons

The autonomic nervous system is concerned with controlling the behaviour of smooth muscle and of many glands throughout the body. Instructions from the CNS are carried to the effector organs not by one motoneuron but by two in series. The axon of the first motoneuron links with the cell body of the second by means of a synapse. The synapses tend to be gathered together in groups forming structures known as the autonomic ganglia.

4. Sensory nerve cells

These each consist of a sense organ which gives rise to an axon. The axon carries nerve impulses to the CNS. Just before the axon reaches the CNS there is the cell body of the sensory nerve cell, usually on a short side branch.

There are two main types of nerve axon, myelinated and unmyelinated. The myelinated ones are larger in diameter and have a sheath of fatty material known as myelin: the unmyelinated ones are smaller and do not have this sheath. As far as function is concerned, the main difference between the two is that nerve impulses travel much faster along myelinated axons than along unmyelinated ones. Many of the most important tracts are therefore made up of myelinated fibres.

LINKS BETWEEN NERVE CELLS

The functioning of the nervous system depends on the ability of nervous impulses to jump from one nerve cell to another. The neurons are arranged so that they come into intimate contact with one another. In sensory nerves the impulses begin at the sensory receptors and travel along the axons to the CNS. There the end of each axon breaks up into minute branches which make contact with the dendrites and cell bodies of other neurons. With motoneurons and most CNS neurons, impulses always begin at the cell bodies and travel out along the axons. The axons of somatic motoneurons make contact with muscle fibres. The places where two nerve fibres come into contact are known as synapses. The membranes of the two fibres do not fuse together and there is a clear gap between them filled with fluid which is outside the nerve cells and in free communication with the rest of the extracellular fluid in the body. The place where a motoneuron makes contact with a muscle cell is a special form of synapse known as a neuromuscular junction: as with other

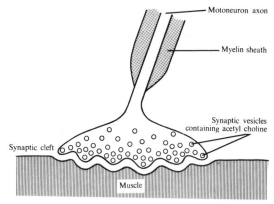

Figure 4.2 The neuromuscular junction.

synapses, there is at this junction a clear gap between the nerve cell and the muscle cell.

OUTLINE OF NERVOUS SYSTEM STRUCTURE

The nervous system consists of three main parts. Two of these, the brain within the skull and the spinal cord within the vertebral column, together form the central nervous system or CNS. The CNS is linked to most of the organs of the body by the third part, the peripheral nerves. Those peripheral nerves which are connected directly to the brain are known as cranial nerves while those connected to the spinal cord are known as spinal nerves.

Like that of a worm, the body of a human being is basically planned on a segmental basis but by birth almost all traces of this segmental origin have disappeared. The clearest remnants of it are seen in the vertebral column and in the spinal cord, which lies in a hole which runs through the vertebrae of which the column is made. There are seven cervical vertebrae, 12 thoracic ones, 5 lumbar ones and 5 sacral ones fused into a single bone, the sacrum. The spinal cord is similarly divided into segments, and corresponding to each vertebra a pair of nerves leaves the spinal cord. The first cervical pair pass out between the first and second cervical vertebrae, the second pair of nerves between the second and third cervical vertebrae and so on. However, the spinal cord is actually shorter than the hole within the vertebral column and it ends near the upper lumbar region. The lower lumbar region contains no spinal cord but only the pairs of nerves passing obliquely down from the cord to the appropriate holes in the vertebrae.

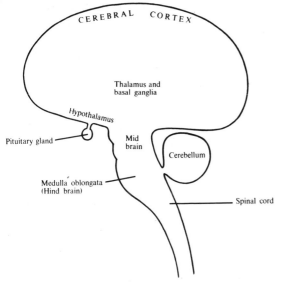

Figure 4.3 An outline of nervous system structure.

Each spinal nerve has two roots connecting it to the spinal cord, one at the back (posterior) and one at the front (anterior). The posterior root consists only of sensory fibres entering the spinal cord. The anterior root consists only of motor fibres leaving the cord on their way to the effector organs. Soon after leaving the spinal cord the two roots fuse and then contain a mixture of sensory and motor fibres.

As mentioned earlier, the motor system is divided into two parts, the somatic system which supplies skeletal muscles and the autonomic system which supplies smooth muscle and many glands. On the whole the somatic system is under voluntary and conscious control while the autonomic system is not, but there are some exceptions to this rule. Structurally, the main difference between the two is that in the somatic system there is only one nerve fibre linking the CNS and the effector. In the autonomic system there are two fibres linked by a synapse. The synapses tend to be gathered together in structures known as ganglia.

The autonomic system itself is further divided into two parts, the sympathetic and the parasympathetic. The sympathetic fibres leave the thoracic and lumbar regions of the spinal cord and their ganglia lie close to the cord in a chain on either side of the vertebral column. One important ganglion, the coeliac, lies in front of the aorta at the root of the coeliac artery. In contrast, the parasympathetic nerves leave the brain and the sacral region of the spinal cord: their ganglia are usually far from the cord in or near the effector organs themselves.

For embryological reasons the brain is usually said to be divided into three parts, the fore-brain, the mid-brain and the hind-brain. The hind-brain is connected directly to the spinal cord and it consists of the medulla and cerebellum. The medulla contains many important control mechanisms which govern sleep, the circulatory system and the respiratory system. The cerebellum is essential for the control of muscular movement.

The mid-brain is small but it receives much information from the eyes and ears. It contains the structure known as the pons.

The fore-brain is the largest of the three sections. It contains the thalamus, an important sensory receiving station, the hypothalamus, vital in the regulation of emotions and behaviour and in the control of many body systems, and the enormous cerebral hemispheres. The great expansion of the cerebral hemispheres is the major aspect of nervous system structure which distinguishes man from other animals. Many of the functional differences between man and animals probably depend on the fore-brain with its cerebral hemispheres.

INTERNAL ORGANISATION OF THE CNS
Within the CNS there are three main types of nervous tissue:

1. The tracts which consist of axons massed together and which go from one part of the CNS to another.

2. Control centres for specific activities such as breathing, control of blood pressure, control of temperature and so on.

3. The poorly understood cortex of the cerebral hemispheres, mainly consisting of a vast mass of neurons with short axons and which is probably responsible for such things as intelligence, consciousness, memory and personality.

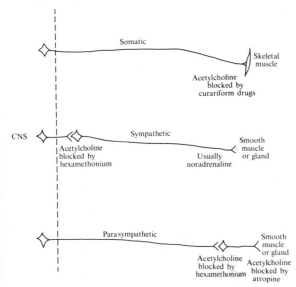

Figure 4.4 The somatic, sympathetic and parasympathetic nervous systems.

There are a few long tracts which go from the brain to the spinal cord or vice versa which it is important to be aware of.

1. Sensory tracts

a. Posterior columns. Sensory nerve fibres enter the spinal cord and, turning headwards, run up the side of the cord on which they entered. In the posterior column nuclei at the top of the cord they synapse with another set of nerve cells whose axons continue the tract which crosses to the other side of the CNS and after synapsing again in the thalamus eventually reaches the cerebral cortex. The posterior columns carry information about light touch, vibration and joint position.

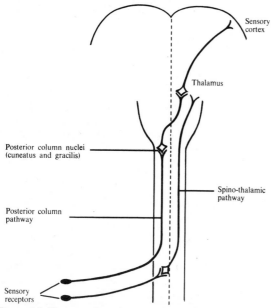

Figure 4.5 The posterior columns and spinothalamic tracts.

b. Spinothalamic tracts. Sensory fibres synapse immediately on entering the spinal cord. The axon of the second nerve cell in the tract then crosses to the other side of the spinal cord and runs up to the thalamus. The spinothalamic tracts carry information about pain, temperature and pressure.

2. Motor tracts

These start in the part of the cerebral hemisphere known as the motor cortex. They cross to the opposite side on their way down through the brain and run down the opposite side of the spinal cord. In the cord they synapse with the motoneurons which leave the CNS and carry the nervous impulses to the muscles. The main motor tract is often called the pyramidal tract.

3. Cerebellar tracts

These go both up and down the spinal cord, linking the cord with the cerebellum on the same side.

It is important to realise that almost all the important connections of the motor and sensory connections of the brain are with the opposite side of the spinal cord and of the body. The main exception is the cerebellum whose connections are with the same side of the spinal cord and body.

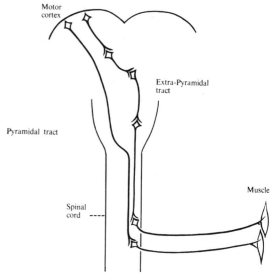

Figure 4.6 The motor tracts.

CEREBROSPINAL FLUID AND LUMBAR PUNCTURE

The brain is a hollow organ containing spaces known as ventricles which are filled with cerebrospinal fluid. There are two lateral ventricles in the cerebral hemispheres which communicate with the third ventricle in the centre of the fore-brain. This in turn communicates through the aqueduct of Sylvius with the fourth ventricle. The latter opens out on to the surface of the brain. The cerebrospinal fluid is formed by structures known as the choroid

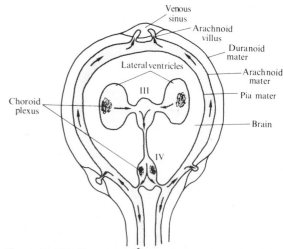

Figure 4.7 The circulation of the cerebrospinal fluid.

villi in the lateral and fourth ventricles. It circulates from the lateral ventricles to the fourth ventricle. Congenital or acquired blocks between the lateral ventricles and the fourth ventricle produce a pressure build up in the third and lateral ventricles leading to the condition known as hydrocephalus in which the head of a fetus or infant swells because of the pressure inside the brain.

After escaping from the fourth ventricle the cerebrospinal fluid (CSF) spreads over the whole surface of the brain and spinal cord. It circulates between the membrane known as the pia mater which covers the brain tightly and the loose, delicate membrane known as the arachnoid mater. Outside the arachnoid is the tough dura mater. There are numerous veins running within the dura mater and protrusions of the arachnoid membrane known as the arachnoid villi stick through the dura into the veins. The CSF passes back into the blood via the arachnoid villi. In the disease known as meningitis when the membranes covering the brain become inflamed, the arachnoid villi become thickened and CSF can no longer pass through them into the blood. But the CSF is still being manufactured by the choroid plexuses and so its pressure rises well above normal levels and may interfere with brain function.

The properties of the CSF are very important in medicine. Four things matter particularly:

1. *Pressure.* This is normally about 10–20 cm of water. It may be raised when some disease stops the normal circulation and absorption of the CSF into the blood. The main conditions which cause a raised pressure after birth are tumours and meningitis.

2. *Blood content.* Normally there is no blood in the CSF but after a haemorrhage in some part of the brain the CSF may become blood-stained.

3. *Glucose content.* Normally the glucose concentration is 50–80 mg/100 ml. However in meningitis, bacteria may use up the glucose and its concentration may be much lower than normal.

4. *Protein content.* This is normally very low indeed but it may be raised in meningitis or in the presence of a cerebral tumour. CSF is normally a watery and crystal fluid. After a haemorrhage it may be reddish or yellowish but still clear. In meningitis it is often cloudy.

It is important for a doctor to be able to sample the CSF reasonably easily. This is usually done by performing a lumbar puncture, and is made possible by the fact that the spinal cord is considerably shorter than the vertebral column and the sack of dura mater and arachnoid mater within the column. There is therefore a fluid-filled space in the lumbar region of the spine below the end of the cord. Under local anaesthetic, a long needle is pushed between the lumbar vertebrae (usually 2 and 3) in the midline. When the needle enters the CSF-filled space, the

pressure of the fluid can be measured and a sample withdrawn.

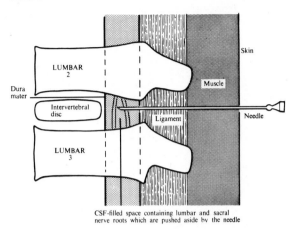

Figure 4.8 The principle of the lumbar puncture.

THE NERVE IMPULSE

The functioning of the nervous system depends upon the electrical disturbances known as nerve impulses (or action potentials) which travel along axons carrying messages from one part of the nervous system to another. Much of our knowledge about nerve impulses has come from the study of squid axons which may be exceptionally large (as much as 1 mm in diameter) and are therefore relatively easy to study.

Suppose that we set up an experiment as shown in fig. 4.9. A single squid axon is mounted in sea water. (Squid

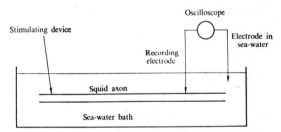

Figure 4.9 The recording of an action potential.

axons are used because they are very large.) At one end there are stimulating electrodes which can deliver an electric shock. At the other end is a very fine electrode (known as a micro-electrode) ready to be pushed into the axon. The micro-electrode is set up so that an oscilloscope can record the electrical potential difference between the micro-electrode and another electrode placed in the sea water bath. While the micro-electrode is outside the axon in the sea water, the oscilloscope of course shows zero potential difference.

Suppose that the electrode is then pushed into the axon. The oscilloscope shows an immediate negative deflection. As with most cells, the inside of the axon is electrically negative to the surrounding fluid. In an axon the inside is usually about 70 mV (1 mV = 1/1000 volt) negative to the outside. This potential difference is known as the resting membrane potential because it is present when the axon is not conducting a nerve impulse.

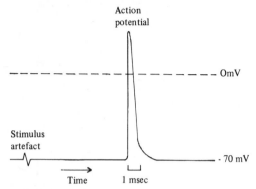

Figure 4.10 An oscilloscope tracing of a nerve impulse.

Suppose that we give to the nerve a single electrical stimulus of low intensity. This stimulus appears as a small deflection on the oscilloscope trace and is known as the stimulus artefact. It is useful because it marks precisely the timing and duration of the stimulus. With a low intensity shock there is nothing to be seen but the stimulus artefact. Suppose the shock strength is gradually increased. Eventually there is a marked disturbance some time after the stimulus artefact. The negative potential difference between the inside and the outside of the axon is momentarily abolished. The inside actually becomes positive relative to the outside. Almost as rapidly, the potential returns to its original level and the whole process is over in about 1·5 milliseconds (msec). This single experiment can give us a great deal of information about the nervous impulse:

1. There is no half way stage between no impulse and a fully developed one. At one stimulus strength there is no impulse while at a fractionally higher one there is a full impulse or action potential as it is known. The shock strength at which the impulse appears is known as the threshold for stimulation. Increasing the shock strength to very high levels does not alter the impulse size. The impulse is either not there at all or it is there in fully developed form. This is an example of what is often called an 'all-or-nothing' phenomenon.

2. During the impulse the inside of the fibre momentarily becomes positive to the outside but the membrane potential then rapidly returns to its resting level.

3. The impulse does not spread down the fibre at the speed of electricity. There is a finite time between the stimulus artefact and the beginning of the action potential. Depending on the size of the fibre and whether or not it is coated with myelin, the speed of conduction may be less than 1 msec or more than 100 msec. Large myelinated fibres conduct impulses most rapidly.

4. The same results are obtained if the positions of the stimulating and recording electrodes are reversed. The axon itself can therefore conduct impulses equally well in either direction.

Although because of its size, much of the early work on nerve conduction was done on the squid axon, it is now known that the findings in squid axons are also in the main true for a mammalian axons as well.

THE REFRACTORY PERIOD

More properties of the nerve axon are revealed if two shocks, one given after the other, are used for testing. Suppose that we measure the threshold required to fire off a single impulse and that we then make the first shock of the pair, A, just strong enough to fire an action potential. Suppose that to start with, the second shock, B, is given a relatively long time (say 10 msec) after A and that again the threshold required to fire an impulse is measured. We should find that the threshold for B was just the same as the threshold for A. Suppose that we then give a series of pairs of shocks and that we steadily bring B closer in time to A. So long as the two were separated by more than about 2 msec we would find that the threshold for B was just the same as the threshold for A. However, if the two shocks were made closer than this, we would find that while the A threshold remained the same, we would have to increase the size of shock B in order to fire off a second impulse. The threshold for shock B would be higher, thus indicating that the nerve fibre was more difficult to excite. As A and B were brought closer together, the threshold for B would rise steeply. Eventually when only about 1 msec separated A and B, the second shock would be unable to fire off a second impulse no matter how powerful it became. Thus for about 1 msec after an action

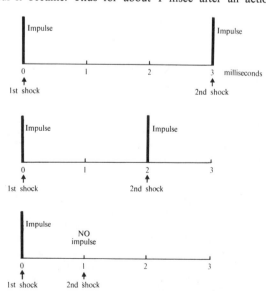

Figure 4.11 The refractory period. When a second shock is given very soon after the first it fails to generate an impulse.

potential has begun, another action potential cannot be fired off no matter how powerful is the second stimulus: this period is known as the absolute refractory period because during it the nerve is completely inexcitable. For another 0·5–1 msec after the absolute refractory period, a second shock must be stronger than usual if it is to start an impulse: this is known as the relative refractory period because, although the nerve can conduct an impulse, it is more difficult than usual to start one off.

An important consequence of these findings is that it is impossible to get two action potentials to add together to make a bigger action potential. A second action potential cannot begin until the first one has been completed.

THE STARTING OF AN IMPULSE

If instead of using the experimental set-up just described, we insert the recording micro-electrode right at the point of stimulation, it is possible to find out something about the way in which an action potential is initiated. Suppose that as before, a single small shock is given to the nerve and that during a series of shocks the strength is steadily increased. The micro-electrode shows that small shocks move the resting membrane potential a few millivolts towards zero. They reduce the potential difference between the inside and the outside of the fibre and are therefore said to depolarise the axon. But with small shocks the small depolarisation produced is not sufficient to fire an impulse and is not conducted along the fibre: the membrane potential returns to its original resting level very quickly. As the shock strength is increased, greater degrees of depolarisation occur. Eventually a shock strength is reached at which the depolarisation develops into a full-sized action potential. This is the threshold

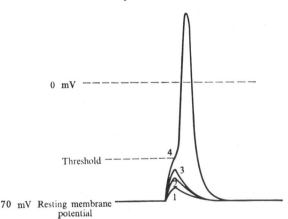

Figure 4.12 Intracellular recording from the point of stimulation showing the effects of four shocks of steadily increasing strength.

shock strength: stimuli weaker than this hold shock strength: stimuli weaker than this do not fire impulses while stronger stimuli always fire impulses.

The essential process in firing a nerve impulse is the depolarisation of the membrane to a certain critical level. In the experiment this was done artificially by means of

an electric shock but as the chapter proceeds we shall see that natural impulses in living animals are generated in a similar way. At synapses the depolarisation is produced by the effect of a chemical transmitter. At sense organs a whole range of stimuli such as light, sound, mechanical deformation and temperature changes can initiate the depolarisation.

THE CONDUCTION OF THE NERVE IMPULSE

Suppose the axon is represented as in fig. 4.13 with the inside of the membrane negative to the outside at rest.

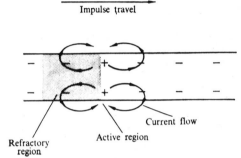

Figure 4.13 The conduction of the nerve impulse.

When an impulse begins, the inside of the fibre at that point momentarily becomes positive. Current can be represented as flowing from positive to negative regions and local electrical circuits will be set up as shown. The current flows away from the active region inside the fibre and back again outside the nerve. The flow inside the fibre from positive to negative areas and then across the membrane depolarises the membrane ahead of the active point. This depolarisation sets up another active focus where again the membrane becomes positive inside and the process is repeated. In this way the impulse passes along the axon. The electricity of course flows both forwards and backwards along the inside of the axon but the impulse does not pass backwards again because the region over which it has just passed is refractory.

THE RESTING MEMBRANE POTENTIAL

Like that of most cells the intracellular fluid inside nerve cells and their axons is rich in potassium and poor in sodium. The extracellular fluid surrounding the cells is rich in sodium and poor in potassium. Some typical figures are shown in table 4.1. The resting membrane

Table 4.1 The compositions of sea water (the fluid outside a squid axon), of the fluid from inside a squid axon, of mammalian extra-cellular fluid and of the fluid from a mammalian muscle fibre. All figures are in milli-equivalents/litre. Squid axons are used in many experiments on nerve conduction because their large size enables them to be handled relatively easily.

	Sea water	Squid axon	Mammalian ECF	Mammalian muscle
Sodium	440	50	140	12
Potassium	22	360	4	155

potential and the action potential are closely tied up with these concentration differentials.

Suppose for a moment that we consider an artificial situation in which a membrane separates a solution of sodium chloride from one of potassium chloride. Three factors will govern the passage of ions across the membrane:

1. The permeability of the membrane to the ion.

2. The concentration gradient across the membrane, ions tending to move from regions where their concentration is high to ones where it is low.

3. The electrical gradient across the membrane, positive ions tending to move towards regions of negative charge and vice versa ('Like charges repel, unlike charges attract'.)

Suppose that the membrane we are considering is impermeable to sodium and chloride but permeable to potassium and that initially we put equal strength

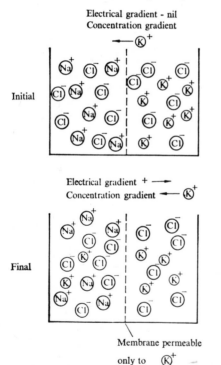

Electrical gradient - nil
Concentration gradient

Initial

Electrical gradient +
Concentration gradient

Final

Membrane permeable
only to K^+

Figure 4.14 The development of a membrane potential (see text).

solutions of sodium chloride and potassium chloride on the two sides. What will happen if these solutions are allowed to come into equilibrium? In order to simplify the situation, osmotic forces will be ignored. Since the other ions cannot cross the membrane we are primarily interested in what happens to the potassium. Since there is no potassium (say) on the left, but a high concentration

on the right, the only force operating initially will be a concentration gradient tending to move potassium from right to left. Initially the only factor limiting this movement is the permeability of the membrane: if the permeability is high the ions will rush across while if it is low they will move more slowly. However, after the first instant electrical forces will begin to operate. Initially both sides are electrically neutral but the potassium ions carry positive charge to the left hand side, leaving an excess of negative charge on the right. This electrical gradient will therefore oppose the further movement of potassium ions. As the movement of ions continues, the potential difference across the membrane will increase until the electrical gradient pushing potassium from left to right is just balanced by the concentration gradient pushing potassium from right to left. Once this equilibrium state is reached, potassium ions will move at equal rates in both directions and there will be no further changes in concentration or electrical potential. The level of potential difference across the membrane at which this happens is known as the equilibrium potential for those particular concentrations.

Nernst made a study of equilibria of this sort and derived an equation to describe them. Using this equation, if the concentration of an ion in two solutions in contact are known, the potential difference which should be set up can be calculated. If the potential difference across a membrane is known, the equation can be used to predict possible concentrations. The equation is as follows:

$$E = K \log \frac{\text{(ionic concentration on one side)}}{\text{(ionic concentration on the other side)}}$$

E is the equilibrium potential. K is a constant depending among other things on the temperature. At 37 °C, K is about 60.

Suppose we apply the equation to a nerve. The membrane is more or less impermeable to chloride and in a simplified treatment the only ions which we need to consider are sodium and potassium. From the known concentrations of sodium and potassium inside and outside the axon, the Nernst equation predicts that the equilibrium potential for potassium will be about 90 mV negative inside while that for sodium will be about 60 mV positive inside. The actual recording resting membrane potential is in the region of 70 mV negative inside.

The potassium equilibrium potential is more negative than the resting potential. In order to hold the observed concentration of potassium ions inside the axon by electrical forces alone the resting potential would have to be −90 mV inside. Since the observed resting potential is only −70 mV inside, there must be some additional force holding potassium inside the axon. Known physico-chemical forces which do not involve the expenditure of metabolic energy are often referred to as 'passive'. Phenomena which cannot be explained by known physico-chemical forces and which require metabolic energy are 'active'. It seems that potassium must be held in the fibre by some active force which operates in addition to the known passive gradients. The word 'pump' is often used as a shorthand name for such an active process. By giving this name to it, however, we have in no way explained what is happening. There is a tendency to give names to things and then to imagine that simply giving a phenomenon a name has in some way explained it. It is important to be on one's guard against using terms in this way.

The sodium equilibrium potential is much further away from the resting potential. The observed concentrations of sodium ions could be accounted for on the basis of passive physico-chemical forces only if the inside of the fibre were 60 mV positive to the outside. As it is, both the concentration and electrical gradients work strongly in the same direction tending to push sodium into the fibre. There must be a powerful sodium 'pump' actively pulling sodium ions in the opposite direction.

The sodium and potassium pumps can account for the observed internal and external concentrations of the two ions. They do not explain why the resting potential should be so very much closer to the potassium equilibrium potential than to the sodium equilibrium potential. The answer to this problem lies in the relative permeabilities of the membrane to sodium and potassium. Measurements of the passive rates of movements of the ions suggest that at rest the membrane is about fifty times more permeable to potassium than to sodium.

The establishment of the resting membrane potential can be best understood if the process is artificially split up into its two major component parts. Suppose that concentration differences similar to those in an axon are set up by sodium potassium pumps operating across a membrane which is totally impermeable to the passive movements of both ions. Once the concentration differences have been established, suppose that the membrane suddenly becomes permeable to passive movements of sodium and potassium in the ratio 1:50. Potassium will therefore rush out down its concentration gradient much more rapidly than sodium will rush into the axon down its concentration gradient. The net effect will be an outward movement of positive ions. This outward movement will continue until the excess of negative charge inside the membrane is sufficient to counter the outward flow of positive charge. Putting it very crudely, because of the membrane's low permeability to sodium, the sodium pump can easily keep sodium out even against a large electrochemical gradient. In contrast, the potassium pump needs the help of a strongly negative internal potential if potassium ions are to be kept inside the fibre.

THE ACTION POTENTIAL

During the action potential, the membrane of an axon momentarily becomes internally positive and approaches the equilibrium potential for sodium. On the basis of what we know about the resting potential, it might be expected that this change in membrane potential could be accounted for by a sudden increase in the permeability of the membrane to sodium. This is in fact what happens. Depolarisation of a nerve results in an increased permeability of the membrane to sodium and sodium tends to enter the fibre. The inward movement of positive ions then tends to cause a further depolarisation. This further depolarisation will bring about a still greater increase in sodium permeability which will allow the sodium ions to move inwards still more rapidly and so on. As a result the membrane potential rapidly approaches the sodium equilibrium potential. Then, for reasons which are still obscure, there is a cut off, the increase in sodium permeability is terminated and the situation returns to its original level.

As mentioned in chapter 2, the action potential is one of the few examples in biology of a positive feedback mechanism. It can occur only because the unstable state of positive feedback is cut off by a return of sodium permeability to normal.

CALCIUM AND TETANY

Normally impulses do not begin spontaneously in nerve cells. They are fired off either by sensory receptors or by an impulse from another nerve cell arriving at a synapse. In the resting state, no impulses begin because the resting membrane potential is stable and does not vary spontaneously. However, the stability of the nerve membrane seems to depend on the presence of precisely the right concentration of calcium ions in the blood and body fluids. The calcium is found in two main forms, as free calcium ions and as calcium bound to proteins in the plasma. The two interact with one another as follows:

$$\text{Calcium ions} + \text{protein} \rightleftharpoons \text{Calcium-protein}$$

It is only the free calcium ions on the left hand side of the equation which are physiologically active in maintaining the stability of nerve membranes. If the concentration of free calcium ions falls too low, nerve axons become unstable and start to fire off many impulses spontaneously. The impulses which arise in motor nerves go to muscles and cause involuntary twitches and spasms. These twitches are particularly evident in the muscles on the inner side of the fore-arm and hand supplied by the ulnar nerve and when severe they give rise to a typical hand position: this is often known as 'accoucheur's hand' because it is similar to the hand position used by a midwife when carrying out a vaginal examination. The facial muscles also often twitch and may sometimes be thrown into spasm if the cheek is tapped where it overlies the facial nerve (Chvostek's sign). The impulses fired off in sensory axons produce tingling sensations. The combination of tingling and muscle twitches is known as tetany and is always caused by a low concentration of free calcium ions in the blood.

Tetany is sometimes caused by a low total calcium concentration, i.e. both the protein-bound and free ionic forms are reduced in amount. This may occur during disease of the parathyroid glands if the output of parathyroid hormone is low. However, this is a rare form of tetany. Much commoner is the situation in which the total amount of calcium in the blood is normal but the concentration of calcium ions is low because they have been bound by protein. Thus more calcium than usual is in the protein-bound form and less is in the form of calcium ions. This situation occurs if the blood becomes too alkaline because alkalinity leads to the binding of calcium by protein. Alkalinity of the blood may sometimes occur because of excessive loss of acid gastric juice following vomiting. Much more frequently it is caused by overbreathing. If you breathe too hard, carbon dioxide is removed more quickly than usual by the lungs and its concentration in the blood falls. Since the acidity of the blood depends to a large extent on carbon dioxide, the fall in carbon dioxide level means a rise in alkalinity. This means that free calcium ions become bound to protein and tetany results. This situation occurs frequently in hysterical and overexcited individuals and also in childbirth. The cure is simple. A bag should be put over the face for a short while so that the patient breathes back the carbon dixide he or she has just breathed out. This quickly brings the blood carbon dioxide level to normal, alkalinity decreases, calcium ions are released from protein and the tetany vanishes.

NEUROMUSCULAR AND SYNAPTIC TRANSMISSION

Impulses jump from one nerve cell to another or from a nerve fibre to an effector organ such as a muscle or gland by a process of chemical transmission. When an impulse reaches the end of an axon it causes the release of a chemical which is stored in the fine branches at the end of the axon. The chemical enters the gap between the axon and the next nerve cell, muscle or gland. It crosses the gap and becomes attached to the membrane of the nerve cell, muscle or gland. On becoming attached, it fires off another nerve impulse or it makes the effector operate.

The process is best understood in the case of the link between a motoneuron and the skeletal muscle fibre which it supplies. This special synapse is often known as the neuromuscular junction. Impulses in the motoneuron

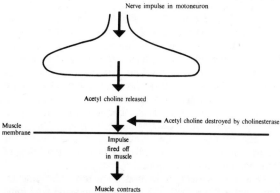

Figure 4.15 The working of the neuromuscular junction.

axon releases a chemical known as acetyl choline. The acetyl choline crosses the narrow gap between the nerve membrane and the muscle membrane and part of the

acetyl choline molecule becomes attached to the muscle: there appear to be special places or receptors on the muscle into which the acetyl choline molecule fits precisely just as the right key fits into a lock. Once the acetyl choline becomes attached it increases the permeability of the membrane to almost all ions. Now if a membrane is permeable to all ions no membrane potential can exist across it and so the membrane potential moves from the region of -70 mV towards zero. This local depolarisation in the region of the neuromuscular junction fires off an impulse which is very similar to a nerve impulse and which spreads rapidly over the surface of the muscle fibre. How this fires off a contraction will be described in the next section.

Almost as soon as the acetyl choline has become attached to the membrane, the end sticking out from the membrane is attacked by an enzyme known as cholinesterase which is found in high concentration at the neuromuscular junction. The acetyl choline is broken down to give an acetyl group and choline and its action is terminated. This destruction allows the muscle membrane to return to its resting state ready for another nerve impulse.

A number of important drugs alter the behaviour of the neuromuscular junction:

1. Drugs like tubocurarine which paralyse the muscle. This group of drugs was originally derived from curare, a South American arrow poison which was used to paralyse hunted animals. Curare is similar enough to acetyl choline to become attached to the membrane but it

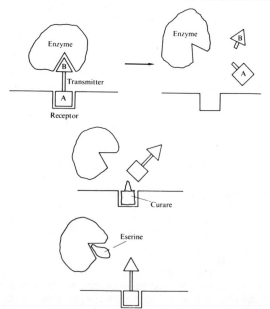

Figure 4.16 Schematic diagram of the action and destruction of acetyl choline and of the modes of action of curare and eserine. The A part of the acetyl choline molecule fits into the muscle receptor while the B part can become attached to the enzyme cholinesterase.

cannot increase the membrane permeability: it is like a key which is almost right but which jams the lock. The muscle receptors therefore become blocked by curare and because the acetyl choline cannot get to them, nerve impulses cease to fire off muscle contractions. Curare cannot be destroyed by cholinesterase and so it has a relatively prolonged action.

2. Drugs like succinyl choline which paralyse the muscle. These drugs are also similar to acetyl choline and become attached to the muscle membrane. They are actually similar enough to be able to 'turn the key' and they depolarise the muscle and cause a contraction. But they cannot be destroyed by the cholinesterase at the neuromuscular junction and so they remain attached to the muscle and prevent it returning to its resting state ready for another impulse. This means that these drugs first produce a twitch and then paralyse the muscle because they cannot be removed. Anyone who has ever watched an anaesthetist put a tube into a patient's trachea has seen the characteristic action of these drugs, which on intravenous injection produce first a series of muscle twitches and then paralysis.

3. Drugs like eserine, neostigmine and prostigmine which block the action of the enzyme cholinesterase. These drugs become attached to the enzyme and thus prevent it from becoming attached to the acetyl choline. The acetyl choline therefore cannot be destroyed in the usual way and so can act for a longer period. To some extent these drugs can reverse the action of those which paralyse the muscle. They are also used in the disease myasthenia gravis. In this disease, although the precise mechanisms are not understood, the patient behaves as though his motor nerves cannot release enough acetyl choline to stimulate muscles properly. By giving anticholinesterase drugs, the action of the acetyl choline that is released can be made more effective and prolonged. This allows muscle power to increase somewhat and the patient may be enabled to live a relatively normal life.

Synaptic transmission in the ganglia of the autonomic nervous system is also quite well understood. At all autonomic ganglia, the chemical transmitter between the two nerve cells is acetyl choline. However, although acetyl choline is the transmitter, the structure of the receptors is slightly different from the structure of the muscle receptors and a slightly different family of drugs act to block the action of acetyl choline. Appropriately enough they are known as the ganglion-blocking agents and they interfere with all the actions of the autonomic nervous system, both sympathetic and parasympathetic.

In the parasympathetic system, the nerves which go from the ganglia to the effector organs all release acetyl choline. Again the receptors are slightly different in structure and these actions of acetyl choline may be blocked by means of a drug known as atropine. Atropine is used in anaesthesia because in response to many anaesthetics, the parasympathetic nerves which supply the bronchi

stimulate the glands there to pour out large amounts of a sticky secretion. This narrows the lung tubes and may predispose to the development of pneumonia after the

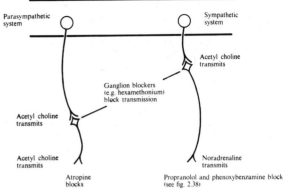

Figure 4.17 Actions of drugs which affect transmission in the autonomic nervous system.

operation. Atropine is therefore usually given some time before an anaesthetic in order to block the actions of these nerves and to prevent the excessive production of secretions.

The nerves which travel from sympathetic ganglia to effectors almost all release a substance known as noradrenaline (norepinephrine). The main exceptions to this rule are the sympathetic nerves which go to the sweat glands and which release acetyl choline.

In the central nervous system, chemical transmission is as yet not so well understood. None of the transmitters present has as yet been identified with absolute certainty although acetyl choline and noradrenaline are present. In the CNS there is an additional complication in that there are transmitters which inhibit the production of nerve impulses as well as those which initiate them. This greatly increases the possibilities for complex interactions.

SKELETAL MUSCLE

There are three main types of muscle, skeletal or striated, smooth or unstriated, and cardiac. The first two will be discussed in this chapter and the third in the chapter on the cardiovascular system.

Skeletal muscle is so-called because it moves the skeleton. It forms the bulk of the soft tissue in the limbs and in the walls of the chest and abdomen. Each bulky muscle is made up of a very large number of long thin cells known as muscle fibres. Muscles are attached to bones by means of very strong structures made of the protein collagen and known as tendons.

Under the microscope skeletal muscle fibres can be seen to have very fine cross stripes, hence the alternative name of striated muscle. The striations are there because of the precise and orderly arrangement within the muscle fibres of the tiny protein fibrils which bring about the actual shortening or contraction. There are two main types of protein fibril, thick ones made of myosin and thin ones, each attached to the fine transverse lines known as Z lines which are made of actin. It is now known that these actin and myosin fibrils are arranged as shown in fig. 4.18, with each thick myosin fibril surrounded by six thin actin fibrils. In the relaxed state

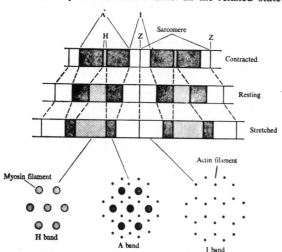

Figure 4.18 The striations of muscle and the changes which they undergo during contraction. The lower figures demonstrate the arrangements of the actin and myosin filaments as seen in transverse section with the aid of the electron microscope.

there is relatively little overlapping of the actin and myosin fibrils. When a contraction occurs, because of a series of events which even now is not well understood, the actin and myosin fibrils become attached and are pulled past one another so shortening the muscle.

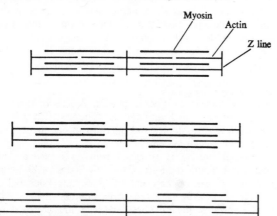

Figure 4.19 During the process of contraction the actin and myosin filaments are believed to slide over one another.

Although it has been intensively studied for many years, there is still much controversy surrounding the biochemistry of muscle contraction and in a book of this type it is not possible to go into details. It will be sufficient to say that the links between actin and myosin which cause the contraction can develop only in the presence of calcium. In the resting muscle there is plenty of calcium but it is held within a system of vesicles known as the sarcoplasmic reticulum which is to be found on either side of the Z line. In a resting muscle the calcium cannot actually get to the actin and myosin. As mentioned in a previous section, the membrane of a muscle fibre behaves in a manner very similar to that of a nerve axon: in fact a muscle fibre might almost be regarded as an axon with a contractile mechanism wrapped up inside it. It is now known that the Z line in fact consists of a series of very fine tubes which cross the fibre and whose membranes are continuous with the outer muscle fibre membrane. When an impulse spreads over the muscle fibre, electrical activity spreads rapidly into the centre of the fibre via the Z lines. This electrical activity then releases a surge of calcium from the sarcoplasmic reticulum. The calcium allows the actin and myosin to interact and a contraction occurs. As soon as the impulse has passed the sarcoplasmic reticulum starts pumping the calcium back in and the contraction ends.

Each muscle fibre is supplied by a nerve fibre which is a branch of a motoneuron axon. All motoneurons split up into several branches and supply several muscle fibres. One motoneuron, its branches, and the muscle fibres it supplies are all together known as a motor unit. In coarse

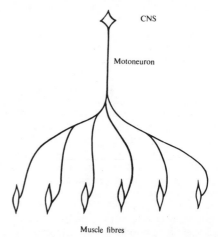

Figure 4.20 A motor unit.

muscles, such as those of the leg, where delicate movements are not required, the motor units are large: a single motoneuron may split up to supply 200 or more muscle fibres. In contrast, in small muscles which carry out fine movements such as those of the fingers or eyes, the motor units are very small and may contain only five or six muscle fibres: this gives the nervous system the ability to control movements much more precisely.

TIME COURSE OF SKELETAL MUSCLE CONTRACTION

The muscle fibre membrane is very similar to the axon membrane. There is a stable resting potential of around −70 mV inside, impulses are conducted rapidly over the surface, each impulse lasts about 1 msec and each is followed by a very brief refractory period. The whole sequence of events of impulse and refractory period is over within less than 2 msec and the muscle fibre membrane is then ready to conduct another impulse.

But the contraction which a single muscle impulse fires off is a much more long drawn out affair. It may take 5–15 msec to reach a peak and may not completely fade away for another 50–100 msec. Furthermore, the contractile mechanism has no refractory period. Two successive contractions are perfectly capable of adding together if the second one starts before the first one has finished.

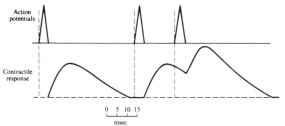

Figure 4.21 Diagram to show how two action potentials which follow one another closely can produce responses which add to produce a larger muscle contraction.

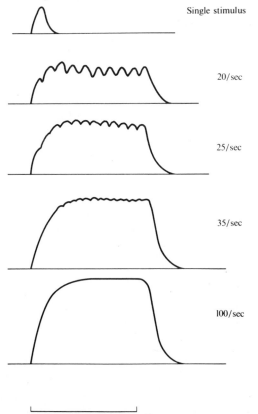

Figure 4.22 The effects of different rates of nerve stimulation on muscle contraction. The response becomes a completely smooth tetanus at frequencies above 50/sec.

Such summation, as it is called, can easily occur because the impulse and its refractory period are over so quickly. The muscle membrane is ready to conduct another action potential long before the contraction following one action potential has reached its peak. A train of nerve impulses less than 10 msec apart will therefore produce a smooth sustained contraction known as a tetanus, with no time for any relaxation following the individual impulses.

MUSCLE LENGTH AND TENSION

If a muscle is held at a series of different lengths and at each length is given a large electric shock sufficient to fire off all the muscle fibres, it is found that the tension which the muscle can develop (the strength of the contraction) depends to a large extent on the starting length of the muscle. When the muscle length is short, the strength of contraction is low. As the muscle is steadily stretched, the tension it can develop when all the fibres contract increases to a maximum level and then, as stretching continues, it falls off again. The peak of the tension curve is usually at about the resting length of the muscle in the intact animal.

The idea that actin and myosin develop tension by becoming linked and sliding past one another has enabled the relationship between length and tension to be plausibly explained. When the muscle is overstretched,

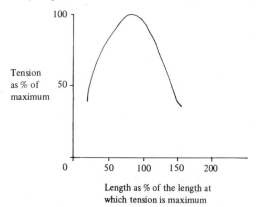

Figure 4.23 The tensions developed when a muscle is held at different lengths and maximally stimulated so that all the fibres in the muscle contract together.

the actin filaments are pulled out from between the myosin ones: there is very little overlapping, few links can be formed and the contraction is weak. When the muscle is very short, the actin filaments are pushed so far inwards that two adjacent ones actually come into contact and interfere with one another. At intermediate lengths there is a maximum of overlapping with a minimum of interference so allowing the muscle to contract with maximum tension.

SMOOTH MUSCLE

Smooth muscle is very important in the body. It is the type of muscle which is found in blood vessels, in the gut, in the lungs, in the ureters and bladder and in the internal genital organs. The fibres themselves are much smaller than those of skeletal muscle and they do not show striations. They contain actin and myosin and the fundamental mechanism of contraction may well be similar in both striated and smooth muscle. However in smooth muscle the contractions appear to be much less efficiently organised and they are slow to develop and to relax. Nevertheless, in most of the situations in which smooth muscle is found, such smooth contractions, far from being a disadvantage, are just what are required.

Many smooth muscle fibres also differ from skeletal muscle fibres in that they have unstable membrane potentials and can contract spontaneously even in the absence of nerve impulses. Simple stretch, for example, often makes a smooth muscle fibre contract. However most smooth muscles do receive nerve fibres which may release acetyl choline (parasympathetic fibres) or noradrenaline (sympathetic fibres). Smooth muscle fibres are also affected by circulating adrenaline (epinephrine) released from the adrenal medulla. In general both acetyl choline and noradrenaline tend to cause smooth muscle contraction. Adrenaline sometimes causes contraction and sometimes relaxation for reasons which will be explained in the section on the autonomic nervous system and the adrenal medulla.

SENSE ORGANS
The function of sense organs is to supply the CNS with information about the exernal environment and also about things which are happening inside the body (the internal environment). Sense organs or receptors which monitor the internal environment are known as proprioceptors. Normally we are not aware that they are functioning although there are exceptions to this rule: for example we are all aware when the bladder is full. Sense organs which monitor the external environment are sometimes known as exteroceptors and we are usually consciously aware of what they are doing.

All sense organs have special endings which are particularly sensitive to some type of stimulus which may be mechanical deformation, temperature, light, some chemical and so on. Usually the receptors are specialised and temperature receptors do not normally respond to light and vice versa. However, very strong stimuli of almost any type may be able to fire off impulses in any type of receptor. The best example of this is the well known fact that a blow on the eye causes the recipient to see stars. The strong mechanical stimulus fires off the light receptors and the brain interprets the impulses as though they had been set off by light.

With each type of receptor, the stimulus which fires it off depolarises the receptor ending: this depolarisation is sometimes known as a generator potential. The depolarisation then fires off an action potential in the nerve fibre leading from the receptor. Receptors differ considerably in the ways in which they respond to a prolonged steady stimulus. Some produce a steady train of impulses which lasts as long as the stimulus lasts and whose frequency is proportional to the intensity of the stimulus. Others produce a burst of impulses when the stimulus is applied but then cease to fire or fire more slowly even though the stimulus continues at its original level. This phenomenon of the falling off of receptor discharge in the face of a constant stimulus is known as adaptation. Most people are familiar with the adaptation of temperature receptors: on first getting under a very hot or a very cold shower the sensation may be very unpleasant but it soon wears off as the receptors adapt.

EXTEROCEPTORS
The most important exteroceptors are:

1. Sense organs in the skin which detect touch, pressure, cold and heat.

2. Taste receptors, especially on the tongue, which monitor the chemical composition of the food. There appear to be four main types for sour, sweet, bitter and metallic tastes.

3. Smell receptors in the nasal passages which monitor the chemical composition of the food. There may be seven different types. The senses of smell and taste are very closely linked. For example many foods 'taste' quite different when our sense of smell is defective as a result of a bad cold. This shows that the sensation which we usually call 'taste' depends on both taste and smell receptors.

4. The eyes and ears which enable us to learn what is happening at some distance away from the body. These senses will be discussed more fully later in the chapter.

PROPRIOCEPTORS
These are vital for the minute by minute regulating processes which keep our bodies running normally and about which we are not usually aware. Some of the more important proprioceptors are:

1. Receptors which measure the blood pressure (baroreceptors). These are found in the walls of many arteries but particularly in the aorta and in the carotid sinus where the internal and external carotids divide.

2. Receptors which monitor the chemical composition of the blood. There are many different types. Important ones are oxygen receptors in the carotid body, carbon dioxide receptors in the brain and carotid body, glucose receptors in the pancreas and brain, pH receptors in the carotid body and receptors for many different hormones, usually in the brain.

3. Receptors which measure the osmotic pressure of the blood. When an animal loses water, the osmotic pressure of its blood may rise: when it gains water the osmotic pressure may fall. Receptors in the part of the brain known as the hypothalamus measure these changes.

4. Receptors in the hypothalamus which measure the temperature of the blood.

5. Receptors in the gut which measure how much it is being stretched by the gut contents and which monitor the chemical composition of the contents.

6. Receptors which monitor the state of the locomotor system. Joint receptors monitor the positions of the joints while Golgi organs in the muscle tendons and muscle spindles among the muscle fibres measure the degree of relaxation and contraction of a muscle. Muscle spindles have been studied in great detail and will be discussed more fully in the next section.

This is by no means a complete list of all the proprioceptors but it does include the most important ones.

MUSCLE SPINDLES

These are exceedingly complex receptors and only a simplified description will be given here. Each muscle spindle consists of several very thin muscle fibres (sometimes known as intrafusal fibres) enclosed within a connective tissue capsule. Several sensory nerve fibres are wrapped around the central region of each muscle spindle fibre. When the main muscle is stretched by some external force, the muscle spindle fibres are stretched and the sensory receptors fire a burst of impulses which then continues but at a lower rate if the stretch is maintained.

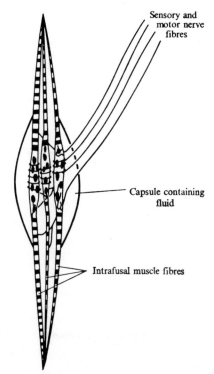

Figure 4.24 The main outlines of muscle spindle structure.

Sensory and motor nerve fibres

Capsule containing fluid

Intrafusal muscle fibres

When the main muscle surrounding the muscle spindle contracts itself, however, tension is actually taken off the muscle spindle and it ceases to fire off impulses.

The muscle spindles are of particular interest because they also receive motor fibres. These are small in diameter and are sometimes called gamma motoneurons to distinguish them from the large alpha motoneurons which supply the main fibres of the muscle. Each muscle spindle fibre usually has two neuromuscular junctions, one on each side of the central region where the sensory fibres are. When the gamma motoneurons fire, the muscle spindle fibres contract and in so doing pull on the central region so firing off impulses in the sensory nerves. In this way the CNS can actually alter the sensitivity of the muscle spindles: even if the main muscle remains in exactly the same position, increased gamma motoneuron activity will increase the activity in the sensory nerves, while reduced gamma activity will relax the spindle fibres and take the tension off the sensory receptors. The precise functional role of this complex mechanism has yet to be worked out but it forms an excellent example of a sense organ whose behaviour is controlled by the central nervous system itself (see chapter 2).

In the tendons of muscles there are other stretch receptors known as the Golgi tendon organs. These differ from the spindles in being much less sensitive and also in responding both to increased contraction of the muscle and to outside stretch of the muscle.

THE EAR

Sound travels through air because the air molecules are set vibrating by the source of the sound. This is most obvious when the sound is generated by a vibrating string as in a guitar but it applies to all types of sound. The ear

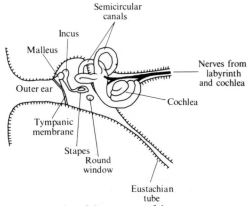

Semicircular canals

Incus

Malleus

Nerves from labyrinth and cochlea

Outer ear

Cochlea

Tympanic membrane

Stapes

Round window

Eustachian tube

Figure 4.25 Outline of the structure of the ear

is a specialised device for picking up these vibrations and for transforming them into nervous impulses. The ear consists of three main parts:

1. *The outer ear*

The outer ear comprises the pinna and a tube leading into the side of the skull known as the external auditory meatus. In animals the pinna may be used for collecting sound waves but in human beings it seems to have lost this function. The meatus contains ceruminous glands which produce wax. The human race seems to be genetically divided into two types: in one the wax is hard and flaky and it falls out easily causing no trouble; in the other the wax is sticky and often blocks the meatus. Accumulation of wax is a common cause of curable deafness.

2. *The middle ear*

This consists of the ear drum, or tympanic membrane and the three bony ossicles known as the incus, malleus and stapes. The stapes is shaped like a stirrup and its foot piece fits tightly into the oval window of the inner ear. The actual sound-detecting mechanism is in the inner ear which is a fluid-filled chamber. The problem which must be solved by the middle ear is to transfer the sound vibrations from the air to the inner ear fluid. Normally sound moves from air to water very inefficiently. The air is much lighter than the water and the air vibrations simply bounce off the surface of the water without setting the water molecules vibrating. The middle ear gets around this problem in two ways:

a. The tympanic membrane which closes off the middle ear from the outer ear is set vibrating by sound. The three bony ossicles transfer these vibrations to the oval window. The tympanic membrane is about fifteen times greater in area than the oval window and so all the energy of the vibrations of the ear drum is concentrated on the window. The effect is rather like that of a drawing pin: it is easy to push a drawing pin into a notice board because all the force which is applied to the broad head of the pin can be concentrated on the tiny point. Thus the vibrations of the oval window are much more forceful than those of the much larger tympanic membrane and are able to set the inner ear fluid vibrating.

b. The ossicles are so arranged that the movements of the stapes at the round window are only about $\frac{3}{4}$ of those of the head of the malleus at the ear drum. In such a linked chain of bones the product of the distance moved multiplied by the force of the movement must be the same at the beginning and the end of the chain. If the stapes moves only $\frac{3}{4}$ of the distance moved by the malleus with each vibration, then the force of movement of the stapes must be 4/3 times that of the malleus.

Because of these two mechanisms, the force of movement of the oval window is about twenty times that of the movement of the ear drum and this enables the vibrations of the air to set the fluid in the inner ear vibrating as well. If you have difficulty in understanding this, think about the drawing pin analogy for a while. If you push the flat head of a drawing pin which has lost its point against a cork notice board you will make little impression. But a normal drawing pin penetrates the cork easily because the force which you apply to the whole head is concentrated on the point which can easily penetrate the cork.

3. *The inner ear*

The inner ear is the place where sound is actually converted into nervous impulses. The organ where this process occurs is known as the cochlea and is shown schematically in fig. 4.26. There are two outer tubes enclosed

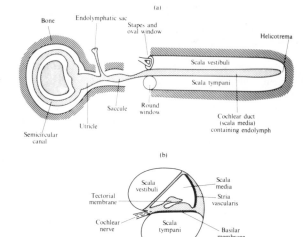

Figure 4.26 a. Schematic outline of the inner ear. For purposes of clarity the cochlea is represented as if it were uncoiled
b. Transverse section of the cochlea.

by bone known as the scala vestibuli and the scala tympani which communicate with one another at the apex of the cochlea. Between these two is a membranous tube known as the scala media which contains the basilar membrane and the organ of Corti. The three tubes are coiled around like a snail shell but in the diagram for simplicity's sake they are shown as though they were straightened out. The hole from the scala vestibuli to the middle ear is closed by the stapes in the oval window while the scala tympani is closed by the membrane across the round window. During vibrations as the oval window is pushed in by the stapes the round window is pushed out and vice versa. The resulting vibrations in the fluid set the basilar membrane vibrating. Low pitched sounds cause the whole membrane to vibrate. As sounds become higher, only part of the membrane vibrates until with the highest sounds only a small length of basilar membrane right next to the oval window is moving.

A simplified diagram of the organ of Corti is shown in fig. 4.27. The basilar membrane carries cells which have

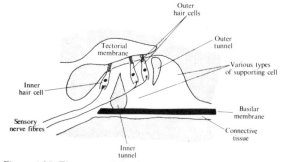

Figure 4.27 The organ of Corti.

hairs sticking out from their upper surfaces. These hairs are embedded in the tectorial membrane. When the basilar membrane vibrates, the tectorial membrane slides over it, pulling on the hairs. This pulling fires off impulses in the nerves which supply the hair cells. Each hair cell has its own nerve fibre. Therefore the brain, by seeing which nerve fibres are firing the most impulses, can tell which part of the basilar membrane is vibrating most: since the part of the membrane which vibrates most depends on the pitch of the sound, the brain can identify what the pitch is. The loudness of the sound depends on just how active the nerve with most activity is.

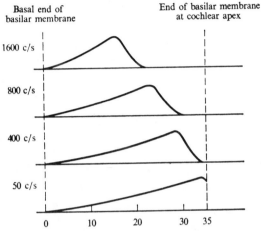

Figure 4.28 Vibrations set up in the basilar membrane by sounds of different frequency.

Deafness
There are essentially two types of deafness, usually known as nerve and conduction deafness. In conduction deafness, the mechanism for conducting the sound to the inner ear via the external auditory meatus, the tympanic membrane, the ossicles and the oval window is defective. The commonest causes of conduction deafness are wax in the outer ear and damage to the ear drum following a middle ear infection. With conduction deafness, sound can still travel to the inner ear via vibrations in the bones of the skull: this is much less effective than the normal mechanism but such patients can be greatly helped by amplification of the sounds by the use of a hearing aid.

With nerve deafness, on the other hand, either the cochlea or the auditory nerve itself is damaged. This means that even if sound conduction to the inner ear is normal, there will be defective hearing. If the auditory nerves are completely destroyed, no hearing aid can help. Overdoses of drugs such as streptomycin and some of the newer antibiotics are not uncommon causes of nerve deafness because they seem to damage the auditory nerve fibres.

The Ear and Balance
The ear is not only the organ of hearing, it is also the organ of balance. The part of the ear concerned with balance is known as the labyrinth and the nerve which

supplies it is the vestibular nerve. The labyrinth contains the semi-circular canals which inform the brain whenever the head is rotated. On each side there are three linked

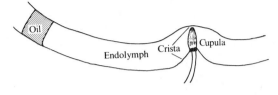

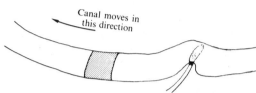

Figure 4.29 The response to rotation of a semi-circular canal. The oil droplet injected into the canal shows that movement of the endolymph lags behind movement of the wall.

semi-circular tubes filled with fluid. Each of the three has at one point a dilatation known as the ampulla. Projecting into the ampulla is a structure rather like a swing door known as the crista. The crista has a rich nerve supply. When the head rotates, the walls of the canals move with the skull, but initially the fluid within the tubes (the endolymph) tends to get left behind: this fluid therefore pushes on the crista, bending it over and firing off a series of nerve impulses. The rate and direction of head movement can be determined by the brain noting which of the canals is generating the most impulses and just how fast the impulses are being fired.

Other parts of the labyrinth are the utricle and saccule. These provide information about the position of the head when it is stationary and also about acceleration. Essentially these organs consist of blobs of chalk mounted on stiff hairs which are in turn connected to nerve fibres. When the head is accelerated, the hairs bend backwards, the degree of bending depending on the speed of the acceleration. When the head is tilted, the hairs also bend and the impulse activity in the nerves increases in proportion to the degree of bending. By assessing this activity the brain can know the position of the head.

THE EYE
Vision can roughly be split up into three separate processes. First, there is a focussing system at the front of the eye consisting of the lens and the cornea. The function of these elements is to produce a sharp image on the retina at the back of the eyeball. Secondly, the retina converts this sharp image into a pattern of nervous impulses. Thirdly these impulses are transmitted to the occipital part of the brain at the back of the head and it is with this part of the brain that we actually 'see'.

Structure

The structure of the eye is shown in fig. 4.30. The eye is divided into two compartments by the ciliary body and

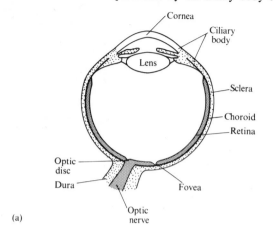

(a)

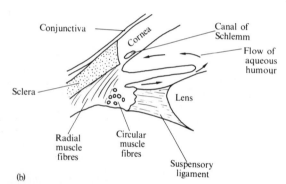

(b)

Figure 4.30 a. Outline of eye structure.
b. Detail of the corneo-scleral angle.

the lens. The posterior compartment is filled with jelly-like vitreous humour while the anterior compartment is filled with watery aqueous humour. The anterior compartment is further sub-divided into anterior and posterior chambers. The posterior chamber is the small space between the lens and the iris, while the anterior chamber is the slightly larger space in front of the iris. The aqueous humour is secreted by the blood vessels of the anterior surface of the ciliary body and of the iris. It passes forwards through the hole in the iris known as the pupil and escapes into the canal of Schlemm. The pressure of the acqueous humour is normally about 15 mmHg but if for any reason drainage into the canal is blocked, the pressure rises. This situation is called glaucoma. It causes intense pain and unless the pressure is quickly brought down the retina may be permanently damaged. There are three basic ways of reducing the pressure:

1. Drugs which constrict the pupil pull the iris tissue away from the canal and allow freer drainage of the fluid.

2. Drugs which reduce the rate of formation of the aqueous humour such as acetazoleamide ('Diamox').

3. Surgical methods of making a tiny hole in the eyeball and so allowing the fluid to drain away.

Focussing

The focussing system consists of two main elements, the cornea and the lens. The cornea is the more important. This is shown by the fact that when the lens becomes opaque in old age or because of disease (cataract), it can be removed. The person is then able to see reasonably well with the aid of glasses. However, the cornea is a fixed lens: it cannot alter its focussing power depending on how far away is the object to be viewed. The adjustable focussing of the eye depends on the lens. The cornea roughly focusses the image so that it falls just behind the retina. The final fine focussing on the retina is done by varying the power of the lens. In healthy young people the lens is able to adjust so that both far distant objects and ones a few inches in front of the nose can be brought into sharp focus. With ageing the focussing power of the lens is reduced and while distant objects can usually be seen normally, ones close to the face cannot be brought into focus. Thus older people almost always require glasses for reading.

The adjustments of the lens depend on the ciliary muscles. The lens itself is in some ways rather like an elastic ball which tends to assume a spherical shape when removed from the body. When in the body there is attached around its margin the tough suspensory ligament which slings the lens from the ciliary body and permanently keeps it partially flattened and less globular than it would normally be. The flatter the lens the weaker it is: the more globular it is the more powerful it is. When

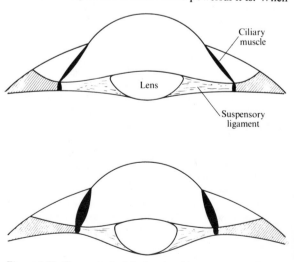

Figure 4.31 The control of the lens by the ciliary muscle. In the upper figure the muscle is relaxed and the suspensory ligament is taut, flattening the lens. In the lower figure the contracted muscle takes the tension off the suspensory ligament and allows the lens to become more globular.

the ciliary muscles contract they pull the suspensory ligaments inward. This takes tension off the lens and allows it to become more globular and powerful. When the ciliary muscles relax, the suspensory ligaments flatten the lens making it weaker. When we are looking at near objects the light rays need to be bent more than when we are looking at distant objects. Therefore, when looking at things close to the eye the lens must be more globular and the ciliary muscles must be working.

In normal people, when the ciliary muscles are relaxed and the lens is flattened, distant objects are in perfect focus. However, if we want to look at an object nearer than about 20 feet away the lens must become more powerful and the ciliary muscles must contract in order to allow it to become more globular. The changes in the eyes which occur when one looks at a nearby object after looking at a distant one are known as accommodation. Accommodation consists of three processes:

1. The contraction of the ciliary muscles to make the lens more powerful.

2. The muscles of the eyeball act to make the eyes look towards the mid-line. This is known as convergence.

3. The pupils become smaller, so covering up the outermost parts of the lens. This is because the central part of the lens focusses much more accurately than the outer parts. When the light rays must be bent considerably in order to achieve focussing, the image is much sharper if only the central part of the lens is used.

The front surface of the cornea must be continually kept moist if it is not to be damaged. This is achieved partly by the tears and partly by oily secretions from glands in the eyelids. Tears are secreted by the lacrimal glands at the lateral edges of the eye sockets. They flow across the eyeball and drain into the nasal passages via the lacrimal duct at the medial side of the eye socket. Tears contain lysozyme, an enzyme which can kill some bacteria. The rate of secretion of tears is increased by stimuli which could potentially damage the eyes such as poisonous gases or dust. It is also increased by emotion.

Focussing Defects

Normal vision in young people is known as emmetropia. If the eye ball is too short or if the cornea and lens are too powerful, distant objects are focussed in front of the eyeball and are always blurred: since there is no way of reducing the power of the focussing system they can never be brought into sharp focus. This condition is known as myopia. Because of the power of their system, people with myopia (short sight) can see objects which are very close to their eyes, almost literally at the end of their noses.

The opposite defect to myopia is known as hypermetropia or long sight. Either the eye is too short or the focussing system is too weak and even distant objects tend to be focussed behind the retina. By accommodation

which increases the power of the lens distant objects can be brought into focus but near ones can never be seen sharply.

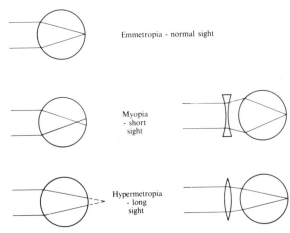

Figure 4.32 Short and long sight and their correction by spectacles.

Fortunately both short and long sight can easily be corrected by the use of appropriate spectacles. With myopia the overall power of the focussing system must be made less and so a diverging (concave) lens must be put in front of the eye. With hypermetropia the overall power of the focussing system must be increased and so a converging (convex) lens is placed in front of the eyes.

The third common focussing defect is known as astigmatism. In this condition the focussing power of the lens

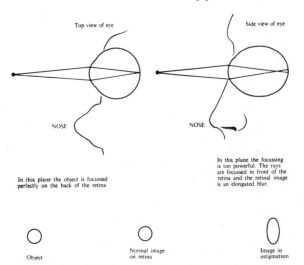

Figure 4.33 Astigmatism.

system (usually of the cornea) is different in different planes. As a result, rays falling in, say, the vertical plane may be focussed precisely on the retina while those falling in the horizontal plane are focussed in front or behind. As a result, round objects tend to appear oval.

Iris

The iris has two main functions. First it constricts in bright light and opens up in darkness. This helps to control the amount of light which falls on the retina. In bright light the iris cuts down the amount of light, while in dim light it opens up as widely as possible in order to let in as much light as possible.

The second function of the iris is to cover up the outer parts of the lens whenever the light is bright enough to permit this. Like all lenses, the focussing of the lens of the human eye tends to be inaccurate when light falls on its edges. The lens functions best when light is passing through its centre. The iris thus helps to ensure that the light falls in the best part of the lens.

The Retina

The retina is the part of the eye which is sensitive to light and which converts the image into a pattern of nervous impulses. The retina contains two types of light receptors, the rods and the cones. The rods are not sensitive to colour but they can pick up very dim light and it is they which function at night. The cones can detect colours but they are much less sensitive than the rods and so function only when the light is relatively bright.

Both rods and cones contain pigments which are broken up when light falls upon them. The substances

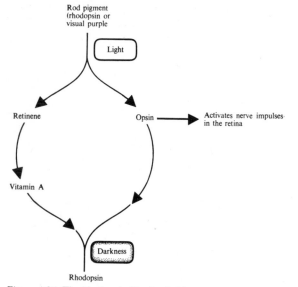

Figure 4.34 The biochemical basis of vision.

produced by this breakdown can then generate nervous impulses. The pigment that is found in the rods is known as visual purple or rhodopsin. Vitamin A is essential for the manufacture of visual purple. In the absence of the vitamin no visual purple can be manufactured and so vision in dim light at night is defective: this is known as 'night blindness'.

The retina has two further layers of cells in addition to the rods and cones. The rods and cones send their impulses to the bipolar cells and the bipolar cells send impulses to the ganglion cells. The axons of the ganglion cells then leave the eyeball via the optic nerve and go to the brain. Paradoxically the rods and cones are on the

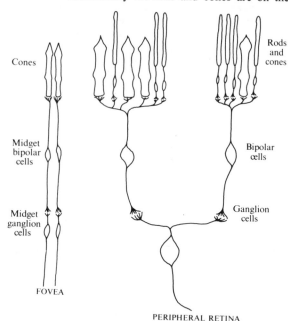

Figure 4.35 The structure of the retina. At the fovea each cone has a 'private line' in the optic nerve. In the outer part of the retina several rods and cones may share the same line.

outermost layer of the retina and so light has to pass through the other cells before it reaches the light sensitive receptors. Only in one area of the retina is this not true. This is the fovea which contains no rods but only cones. In the fovea the ganglion and bipolar cells are pushed to one side so that light can fall directly on the cones. Visual acuity is highest when the image falls on the fovea and so when you look at an object the eyes automatically move so that the image of the object falls directly on to the fovea. This is why only the very centre of an area at which you are looking is seen clearly and why things at the edge of the field of vision tend to be hazy.

All the ganglion cell axons are gathered together in the optic nerve which leaves through the back of the retina. At the point where the optic nerve leaves the eyeball there are no rods or cones and so if light falls on this area it cannot be detected. This is known as the blind spot.

Colour Vision

Light can be regarded as consisting of waves. The wavelength of light, the distance from the peak of one wave to the peak of the next determines the colour which we see. The eye as a whole can detect light waves whose wavelengths range from about 400 to about 700 mu (1 mu is one thousand millionth of a metre). Visual purple can be broken down by light of any wavelength within this range and so the rods can see only one 'colour'. The cones are quite different. There are three distinct types, each of which is specialised to detect light of a much more limited range of wavelengths. Blue light has short wavelengths near the 400 mu end. Green light has medium length wavelengths in the 500–600 mu range. Red light has relative long wavelengths in the 600–700 mu range. The three types of cones are each specialised to detect light of a particular type. The pigment in the 'blue' cones is broken down by light in the blue wavelength range and so the blue cones will fire impulses when

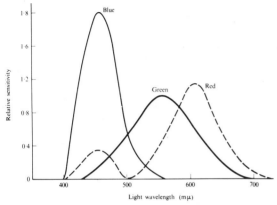

Figure 4.36 The three types of cone which are believed to be important in colour vision. Each is particularly sensitive to light of a particular wavelength.

blue light falls upon them but not when red light reaches them. The 'red' cones have a pigment which is broken down by light of the red wavelength range but not by blue light. They will therefore fire impulses when red light falls on them. The 'green' cones have a pigment which is intermediate between the two.

If you have ever taken any interest in theatre lighting you will know that you can produce light of any colour, including white, by using three lamps, one red, one green and one blue. When all three lamps are shining on a spot with equal intensity, the whole range of visible wavelengths is represented and the result is white. All the other colours may be made by altering the intensities of the three lights. Colour vision works in a roughly similar way. If only red light falls on the retina only the 'red' cones will fire impulses. Nothing will happen in the green and blue cones and the brain can therefore 'know' that the light falling on the eye must be red. Similarly, if pure green or pure blue light falls on the retina, only the relevant cones will be stimulated and the brain can know what colour the light must be. If white light, which contains all the wavelengths from 400 to 700 mu falls on the retina, all three cones will be firing equally. The brain will note that all the cones are firing, that the light must therefore be white. Other colours are formed by different patterns of cone firing. A yellow sensation, for example, occurs when the green cones are firing rapidly and the red cones are firing moderately. This is because yellow light has a wavelength between green and red light but which is closer to green. An orange light on the other

hand produces rapid firing of the red cones and moderate firing of the green cones. This is because orange light has a wavelength which is closer to red than to green. Thus by looking at the pattern of firing in the different types of cone the brain can know the colour of light which is falling on the retina.

Dark Adaptation

When you go from the light outside into a darkened room, you are all familiar with the sensation of being quite unable to see anything because there is not enough light reaching your eyes. Gradually, as you become accustomed to the gloom, you can see more and more: your vision continues to improve for about 25 min after which no further improvement takes place. No matter how long you stay in the dark you cannot see colours, although you may be able to see things quite clearly in shades of grey. This process is known as dark adaptation.

The sequence of events in dark adaptation depends on two facts. First, in ordinary daylight or in bright artificial light, the brightness of the light is sufficient to break down all the visual purple which is immediately available. Since, if there is no intact visual purple to be broken down, the rods cannot detect light, the rods are functionless in bright conditions. Secondly, the cone pigments are much less sensitive to visual purple. They are more resistant to being broken down and they cannot detect light whose intensity is below a critical level. So in the bright surroundings the rods are functionless because the bright light has broken down the available visual purple and you rely entirely on cone vision. When you go into a dark room, the cones cannot work because the light intensity is too low while the rods cannot work because they do not contain enough available visual purple. You are therefore temporarily blind. But as soon as the bright light is removed, visual purple begins to be manufactured again in the rods and as its concentration rises so the rods again become sensitive to light. The process continues for about 25 min, when the maximum sensitivity is reached.

An interesting phenomenon which you may have noticed is that at night when you try to look directly at an object such as a star it seems to disappear. When you look a little to one side, the object appears again. This occurs because when you look directly at something, the image falls on the fovea where there are only cones: since the cones are functionless in dim light at night you cannot see objects whose images fall on the fovea. But if you look to one side the image falls not on the fovea but on a part of the retina which contains rods and you can see the object again. When you are trying to see something at night it therefore pays not to look directly at it.

Colour Blindness

Some people cannot distinguish between colours in the normal way. This is because instead of having three types of cone they have only one or two. This means that it is difficult or impossible for such individuals to distinguish between certain colours: the difficulties vary depending on the cones which are missing but the commonest one is a difficulty in distinguishing between red and green.

The normal function of cones seems to depend on genes which are carried on the X chromosome. In females there are two X chromosomes but in males there

X   X   X   Y

Normal female   Normal male

Normal female   Colour-blind male

The gene which prevents red-green colour blindness is on the X chromosome. Both X chromosomes in the female must be abnormal if a female is to be colour-blind. A single abnormality in the only X chromosome causes colour-blindness in males.

Red-green colour-blind female

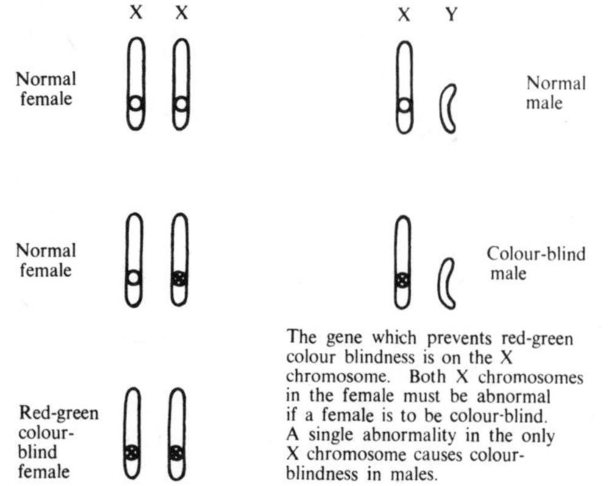

Figure 4.37 The inheritance of red-green colour blindness.

is only one. In females it is possible for one X chromosome to be defective, but so long as the other X chromosome is normal, colour vision will be normal. In females the common forms of colour blindness will occur only if both X chromosomes are defective. In males, however, with only one X chromosome, there is no such safety factor. If that single chromosome is defective, colour blindness will occur. This explains why the common form of red-green colour blindness is so much commoner in males than in females.

Eye Movements and Visual Fields
Normally the two eyes move together so that the image of a particular object falls on the fovea of each eye simultaneously. In other words the two eyes look at the same thing and the two separate images are fused into one by processes which occur in the brain. If the two eyes do not look at the same point at the same time, the person is said to have strabismus or a squint. Double vision may occur but is often absent because the brain totally ignores the information coming from the abnormal eye and takes notice only of the normal one.

Because of the nose and the margins of the eye sockets, the area seen by one eye is not exactly the same as that seen by the other eye. The area seen by one eye is known as the visual field of that eye. You can easily confirm that the two visual fields are slightly different by looking straight ahead and shutting each eye in turn. The assessment of visual fields is an important part of the examination of a patient because it may reveal damage to the eyes, to the optic nerves or to the brain.

In order to understand why the visual fields should be tested it is important to know the pathways which the nerves take from the eyes to the occipital cortex. These are shown in fig. 4.38. The first thing to note is that the light rays cross over in the focussing system so that the image of objects on the right side of the body falls on the left side of the eye and vice versa. Secondly, the nerve

fibres from the outer (lateral) half of each retina go to the same side of the brain while those from the inner (medial) half of each retina cross over in the optic chiasma and go

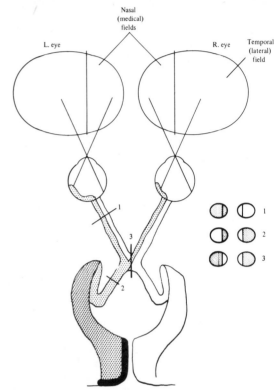

Figure 4.38 The visual pathways from the eye to the occipital visual cortex and the effects on the visual fields of cuts at various points.

to the opposite side of the brain. In fig. 4.38 various cuts are shown and the explanation of the damage which each one causes is given below.

1. This cut severs the optic nerve on one side before the optic chiasma. Thus all the fibres coming from one eye will be destroyed and total blindness will occur on that side. The visual field of the other eye will be normal.

2. This cuts the optic pathway behind the chiasma. It therefore destroys the fibres carrying information from the outer side of the retina on the same side as the cut and those from the inner side of the retina on the opposite side to the cut. Thus if the cut is on the right side, all the fibres coming from the right sides of both eyes will be destroyed. Since the right side of each eye sees objects on the left side of the body, in each eye, the left side of the visual field will be missing. If the cut is on the left side, the right side of each field will be defective.

3. This cuts the optic chiasma. Such damage could occur naturally as a result of the growth of a tumour of the pituitary gland which lies just below the chiasma. The fibres

from the inner side of each retina will be destroyed. In the left eye the right side of the retina will be affected and so the eye will be unable to see objects on the left side of the body. Similarly the right eye will be unable to see objects on the right side of the body. When both eyes are open each eye will tend to compensate for the other's defects.

REFLEXES

All actions in some way depend on sensory information received by the CNS. The information may have been stored in the brain as a result of past experience as when a pianist plays a piece of music from memory. On the other hand the information may have been received immediately before the motor act as when someone playing tennis moves to where she sees the ball is going. In many cases we are aware of what is going on and we consciously direct the motor action. In other cases, although we may be aware of what is happening, the motor response occurs involuntarily without our conscious direction, as when a hand is pulled away from an unexpectedly hot plate. In still other situations we neither direct the motor response consciously, nor are aware that it is happening, as when the body regulates arterial pressure in order to cope with a fall or a rise in pressure. The last two examples are both known as reflexes. We can say that a reflex occurs whenever a sensory stimulus leads to a motor (effector) response without the intervention of conscious, voluntary control. In order for a reflex to occur, five essential components must be present:

1. *A sensory receptor* for detecting that the stimulus has occurred.

2. *A sensory nerve fibre* for carrying the information to the CNS.

3. *A control centre* in the CNS which collects the sensory information and decides on appropriate motor action: at its simplest this centre may consist of a single synapse.

4. *A motor nerve* which carries information from the CNS to the appropriate effector.

5. *An effector organ* (usually a muscle or a gland) which carries out the response.

Reflexes vary a great deal in their complexity. The simplest can occur even after damage to the spinal cord has cut it off from the rest of the CNS. Such simple reflexes are known as spinal reflexes and have naturally been studied most in experimental animals. Nevertheless, they can also be clearly seen in patients whose spinal cord has been damaged by accident or by disease. The most important spinal reflexes are:

1. *The stretch reflex or tendon jerk.* This is best illustrated by the familiar response to tapping the front of the knee. The leg is allowed to hang freely and a sharp tap is given to the tendon which stretches from the knee cap to the tibia. This gives a brief stretch to the quadriceps muscle on top of the thigh. The stretch stimulates sensory receptors known as muscle spindles which consist of

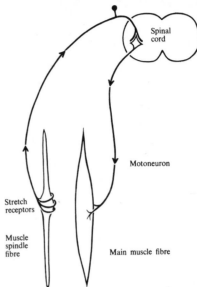

Figure 4.39 A muscle spindle and the pathway involved in the stretch reflex.

specialised muscle fibres with sensory nerves wrapped around them. The sensory fibres carry the impulses up to the spinal cord. In the spinal cord the sensory fibres synapse with motoneurons which go back to the muscle which was stretched. The impulses going along these motoneurons cause the muscle to contract. Thus a brief stretch of any muscle is always normally followed by a brief contraction of the stretched muscle. One possible function of the stretch reflex will be discussed later under the heading of posture.

2. *The flexion reflex.* This is the response of a limb to a painful stimulus. Whenever any potentially damaging stimulus is applied to a limb, it is detected by pain receptors which send the information to the spinal cord. This leads at once to a coordinated movement of the limb in which the flexor muscles contract and the extensors relax so pulling the limb away from the danger zone.

3. *The crossed extension reflex.* This is closely related to the flexion reflex. If you tread on something sharp and the flexion reflex pulls your leg up quickly, you will fall unless you are supported by your other leg. In fact you do not usually fall because the extensor muscles in the unstimulated leg contract, so making the limb into a firm pillar on which you can stand. Thus a painful stimulus to one leg causes extension of the other leg and this is known as the crossed extension reflex.

4. *The response to firmly stroking the underside of the foot with a blunt point.* In a normal individual this causes curling up of the toes and in particular of the great toe. In

someone whose motor pathway from the motor cortex to the spinal cord has been damaged, the great toe, instead of flexing downwards extends upwards. This is known as the Babinski response and is usually an indication of severe damage either to the spinal cord or to the motor tracts in the brain. The only exception to this rule is that the Babinski response is normal during the first 12–18 months of life before the spinal cord tracts have received their normal coats of myelin. The process of the myelination of the motor tracts continues for quite a long period after birth and is associated with the progressive development of the infant's motor behaviour.

5. *Micturition* (urination) in response to a full bladder. This occurs automatically in infants and in those whose spinal cord has been damaged above the lumbar region. In normal individuals the micturition reflex is, of course, brought under voluntary control.

6. *The defaecation reflex* is similar to the micturition reflex.

7. *Erection of the penis and ejaculation of sperm*. This can be brought about after spinal cord section by stimulation of the glans of the penis. The person being stimulated cannot feel what is happening. However, this reflex enables paraplegic patients (ones whose spinal cord has been cut across) to father children. The sperm can be collected and the wife fertilised by artificial insemination.

Many other reflexes operate to keep the body running smoothly but they require the participation of parts of the brain above the level of the spinal cord. Important examples of these more complex reflexes are:

1. The responses of the respiratory system to oxygen lack and carbon dioxide excess(chapter 8).

2. The responses of the cardiovascular system which help to maintain the constancy of the arterial pressure (chapter 7).

3. The control of the output of aldosterone and anti-diuretic hormone. These hormones modify the behaviour of the kidneys and help to keep the composition of the body fluids constant (chapter 9).

4. The reflexes involved in the maintenance of posture which are described later in this section.

CONDITIONED REFLEXES

All the reflexes so far described are inborn. They are possessed by every normal individual, develop automatically and do not have to be learned. Conditioned reflexes, while also being involuntary motor acts in response to a sensory stimulus, are different in that they are acquired by experience. They were first clearly demonstrated by the Russian scientist Pavlov, using dogs. Dogs, like most animals, respond to the sight and

smell of food by pouring out saliva. Pavlov devised a method of collecting this saliva and measuring its quantity. In a group of dogs, every time he provided food he simultaneously rang a bell. After he had given the food and rung the bell together a number of times, he then rang the bell without giving the food. The dogs still poured out saliva. They had come to associate the bell with the food and therefore salivated even when no food was provided: they thought that the bell was an indication that food would be coming. Daily life is full of examples of such conditioned reflexes. For example, small children do not naturally refrain from running into the road when they hear the sound of a car engine. Only as they grow and gradually begin to associate cars with danger do they automatically jump off the road when they hear a car coming. If you think a little you will find many other examples of acquired conditioned reflexes in your own life. They are reflexes in which a stimulus is not naturally associated with an involuntary response but in which it becomes so as a result of experience.

SENSORY PATHWAYS IN THE CNS

We are not aware of much of the information which is continually being collected by our sense organs. Only a small proportion of the information reaches the level of consciousness. When sensory information is consciously appreciated it is said to be perceived. Perception is the being aware of a particular sensory event.

Sensory information is carried from the receptors to the brain and spinal cord by sensory nerves. In the CNS these nerves make contact with other nerve cells which carry the impulses up to the thalamus and to the cerebral cortex and to the cerebellum. There is some evidence that crude sensations are perceived at the thalamus but that their precise localisation and identification requires the participation of the cerebral cortex. For example, after destruction of the cerebral cortex which leaves the thalamus intact, it is possible to be aware that a painful stimulus has been applied to the body but it is impossible to localise it precisely.

The routes via which the most important types of sensory information reach the brain are outlined in fig. 4.5.

1. Information about pain, temperature and pressure on the skin below the level of the neck. The sensory fibres enter the spinal cord and synapse immediately. The fibres of the spinothalamic tract then cross to the other side of the spinal cord and carry the information up to the thalamus. These types of information do not appear to reach the cerebral cortex.

2. Information from below the neck about light touch to the skin, vibration and the position of the joints. The fibres from receptors enter the spinal cord and without synapsing travel up in the posterior columns of the cord to the posterior column nuclei at the junction of brain and cord. There they synapse and another set of fibres carries the impulses up to the thalamus of the opposite side. A third set of nerve cells then transmits the information to the sensory cortex.

3. Information from the skin of the head of all types enters the brain via the trigeminal nerve and is carried to the thalamus of the opposite side. Light touch and vibration information is then carried up to the sensory cortex.

4. Information about sound enters the brain via the auditory nerve. Some is sent to the cerebral cortex on the same side and some to the opposite side. This means that each half of the brain receives information from both ears.

5. Information from the eyes was dealt with in an earlier section.

SENSORY AREAS OF THE CEREBRAL CORTEX
While vague sensations of light, sound, pain, touch and so on may probably be appreciated at the thalamic level, precise identification and localisation of stimuli requires the participation of the cerebral cortex. Three main areas

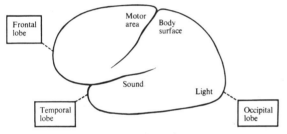

Figure 4.40 The sensory areas of the cerebral cortex.

are involved. Dividing the cortex into anterior and posterior parts is a deep central fissure (sulcus). Immediately behind this is the area devoted to analysing information from the body surface and the joints. The body surface is not evenly and proportionally represented. For instance, the areas of the cortex devoted to the hands and to the lips are out of all proportion to the skin areas involved and are enormous compared to the cortical areas devoted to the trunk and to the legs. This is presumably because the hands and lips are exposed to a much greater variety of stimulation than the trunk and legs. The whole area of cortex which deals with the skin surface is sometimes known as the somatic sensory cortex. If it is destroyed on one side, sensation on the opposite side of the body is lost.

The part of the cortex which deals with vision, the visual cortex, is right at the back of the brain in the occipital region. The left occipital cortex receives information from the left side of each retina. The left side of each retina receives light from the right hand side of the body. Therefore if the left side of the occipital cortex is destroyed, a person looking straight ahead will be unable to see objects on his right.

The auditory cortex lies below the somatic sensory cortex in the temporal lobe of the brain. Each part of the auditory cortex receives information about sound from both ears and so damage to one side causes little if any hearing loss.

THE MOTOR CORTEX
The voluntary control of muscular activity depends to a large extent on the motor cortex which lies in front of the central sulcus. It is arranged in much the same way as the sensory cortex. Large areas are devoted to the hands and face which carry out a wide range of delicate movements, while much smaller areas are devoted to the trunks and legs which carry out a much smaller range of relatively crude and stereotyped movements. Instructions about

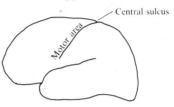

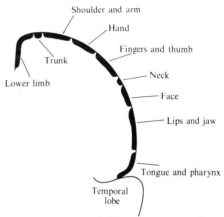

Figure 4.41 The position of the motor cortex and a coronal section through it showing the relative sizes of the areas which deal with muscles in different parts of the body.

voluntary movements are sent down from the motor cortex by two systems of fibres, the pyramidal and the extra-pyramidal tracts. In the pyramidal tract, single continuous fibres go right from the motor cortex down to the motoneurons of the spinal cord. In the extra-pyramidal tract the information is carried by a chain of shorter neurons. Both tracts cross in the brain to the opposite side and so each half of the motor cortex deals with the opposite side of the body. Damage to the motor cortex therefore causes paralysis on the opposite side of the body.

OTHER ASPECTS OF CEREBRAL CORTICAL FUNCTION
The brain has many other functions apart from the ones which have so far been discussed. Some, such as memory, we are very far indeed from understanding and there is little point in dealing with them here. Other functions which we are just beginning to understand are sleep and speech and these will be discussed in this section.

Sleep

The electrical activity of the tens of millions of nerve cells of which the cerebral cortex is composed can be recorded through the skull. Sensitive electrodes placed at many points on the surface of the scalp can pick up this activity and feed it into a machine known as an electroencephalograph (EEG) or electrocorticograph. This amplifies the electrical signals and records them on moving paper.

When a person is awake and alert, the oscillations of electrical activity are very fast and irregular. When a person becomes drowsy and closes his eyes, the waves become slower and more coordinated giving the so-called alpha rhythm. On going to sleep the further changes shown in fig. 4.42 can be seen: large slow waves alternate

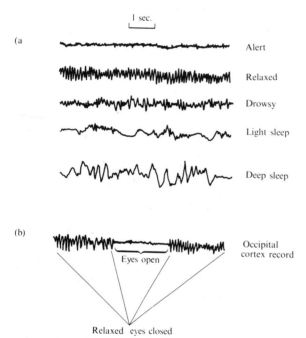

Figure 4.42 a. Some typical EEGs showing the transition from the alert state to deep sleep.

b. Opening the eyes abolishes the alpha rhythm seen in most normal adults when resting quietly with the eyes closed.

with bursts of high frequency activity known as sleep spindles. The basic rhythm seen in a relaxed individual varies with age. In infancy slow delta rhythms of less than 4/sec dominate. During childhood the theta rhythms (4–8/sec) predominate, gradually giving way to the mature alpha pattern (8–14/sec) after puberty. Apart from its use in experimental situations as an indicator of sleep, the EEG is of importance in four main situations:

1. In certain psychiatric behavioural disorders, the theta rhythms persist into adult life reflecting the immaturity of the personality of the patient.

2. In epilepsy, abnormal electrical discharges may be seen and the site of their origin indentified.

3. Cerebral tumours, abscesses or other abnormalities within the skull may produce an abnormal EEG pattern.

4. The EEG may be used to try to decide whether a person's cortex is dead or alive. This may be important after injury or drug overdose when the heart may be beating normally but the brain (and therefore the person) may be dead.

The state of sleep seems to be controlled by a system of nerve cells in the core of the mid-brain and hindbrain known as the reticular formation or the reticular activating system. The reticular formation has connections

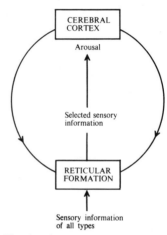

Figure 4.43 The function and connections of the reticular formation.

with the whole cerebral cortex and its activity determines whether the person will be awake and conscious or asleep and unconscious. Experiments with animals have shown that destruction of the reticular formation produces a permanent state of sleep, while stimulation of it by means of electrodes implanted into it produces a permanently awake state. In humans, haemorrhage into the mid-brain (pontine haemorrhage) damages the reticular formation and causes a state of deep unconsciousness from which the patient cannot be roused. Many sleeping pills, and in particular the barbiturates, seem to act by suppressing the activity of the reticular formation.

The reticular formation appears to work by collecting information from two major sources and using that information to regulate the activity of the cortex. The sources are:

1. The situation in the environment as determined by the activity of peripheral sensory organs. Everyone acknowledges the importance of this by trying to reduce sensory activity as much as possible before going to sleep. We switch off the radio and the light and try to ensure that

the bed is neither too cold nor too hot. This cuts down the sensory information pouring into the reticular formation and reduces its activity.

2. The state of cortical activity. Again everyone is aware of the importance of this. Impending importance events such as examinations, an exciting date or marriage keep us thinking even if the environmental situation is ideal for sleep. The cortex remains active and the reticular formation will not allow us to go to sleep.

The reticular formation is more than a simple collector of information. It also assesses the significance of the information received. For example, if you live by a railway line you soon cease to be woken up by the trains running at night. Your reticular formation receives the information about the train from your ears in the usual way. However, it soon learns that this information is of no importance and so does not bother to rouse the cortex and wake you up. Yet if your bedroom door creaks while a train is thundering past you are awake in an instant. The reticular formation has decided that the small sound of the creaking door may be very important and has aroused the cortex to activity. The reticular formation therefore receives all the information coming from sensory organs but it acts to arouse the cortex only if the information seems important.

Speech
The ability to speak involves a remarkably complex series of processes. First, it is important to understand what words are and what they mean. Second, the words which are appropriate to a particular situation must be selected. Third, instructions must be sent to the muscles of the chest, the larynx and the mouth in order that they may produce the required sounds. It is not surprising that we are only just beginning to understand how the brain deals with speech.

The use of words is clearly closely involved with the hearing of sound and with the seeing of the printed page. It also involves the control of muscles of the chest, larynx, throat and mouth. It therefore seems sensible that the area of the cerebral cortex which deals with speech should lie between the three areas which deal with sight, with hearing and with muscular activity. It has been found that in most individuals it is the dominant hemisphere which is most important in speech. Thus, in right handed individuals in whom the left hemisphere is dominant, speech depends primarily on the function of the left hemisphere. In left handed individuals speech depends on the right hemisphere. There is some evidence that many people who stammer are naturally left handed but in childhood were forced to become right handed by parents or teachers who mistakenly thought that right handedness was somehow better. This attempt to change the dominant hemisphere by outside action obviously must have serious consequences for the organisation of the brain, leading to problems in speaking.

There are three relatively simple disorders which lead to problems in word use. Blindness makes it impossible to read with the eyes although it may be possible so to develop the sense of touch that the Braille symbols can be interpreted as letters. Deafness makes it impossible to hear sounds and in the development of speech is a much more important defect than blindness. Deaf children cannot hear speech and cannot naturally learn to speak. Unless they receive special education they remain dumb. Dysarthria is an inability to formulate sounds properly because of some damage to the motor cortex or motor pathways which interfere with the normal control of muscular activity. In none of these situations is there any real difficulty in the understanding and choosing of words.

Aphasia is a much more complex problem: it is a condition which results from damage to the speech area in the dominant cerebral hemisphere often as a result of a cerebrovascular accident or stroke. The patient has difficulty in understanding words, in arranging them in their correct order and in selecting the ones which he wants to use. As can be imagined this is a very distressing condition since it makes it extremely difficult for anyone to communicate with the patient.

HYPOTHALAMUS
The hypothalamus, as its name implies, lies beneath the thalamus at the base of the forebrain. It is a region which is currently being much investigated because it seems to be the site involved in much of emotional and sexual behaviour. In animals, stimulation of the hypothalamus by electric currents can cause uncontrollable rage or perfect calm, fear or aggression, eagerness to mate or lack of interest in sex. As yet we know very little about the significance of these things in humans. Certainly some tumours of the hypothalamus can result in very strange human behaviour but such tumours are rare and difficult to study.

The hypothalamus is also important because it controls the behaviour of the pituitary gland, the so-called 'master gland' in the body. This is discussed fully in the next chapter.

POSTURE AND MOVEMENT. THE CEREBELLUM
One of the most important functions of the nervous system is the contol of muscular activity. But it is a mistake to think that this control depends only on the motor side of the system. The sensory side is just as important. This is clearly illustrated by considering the difference between playing tennis with your eyes shut and your eyes open. When the motor system is deprived of sensory information it becomes almost useless. Only if it receives a continuous supply of reliable sensory information can the motor system work safely and effectively.

A major part of the work of the motor system concerns the maintenance of posture, particularly when you are standing up. The important sensory receptors which provide the brain with information about posture are as follows:

Figure 4.44 Outline of the systems involved in the control of posture.

1. The eyes. These provide a constant stream of information about the position of the body, of the head, of the limbs and about objects in the environment.

2. The labyrinths in the inner ear. These provide information about the position and the movement of the head.

3. Receptors in the joints of the neck. These supply information about the position of the head in relation to that of the body.

4. Receptors in other joints. These provide information about the position of the limbs. Conscious knowledge of limb position depends on these joint receptors and not on receptors in muscles.

5. Receptors in the muscles known as muscle spindles and Golgi tendon organs. These provide the CNS with information about muscle stretch and contraction but we are not consciously aware of it.

6. Skin receptors which provide information as to which part of the skin is taking the weight of the body.

All this information is made available to the motor areas of the brain and is continually put to use as the CNS directs muscular activity. Of all the receptors, the most important are the eyes. However, damage to any one set of receptors may lead to disturbances in posture, especially when the eyes cannot be used properly as in blind people or in the dark.

The Cerebellum
One of the most formidable problems faced by the nervous system is that of precisely controlling the strength of muscular contraction in the face of the very variable amounts of force which must be used. For example, how does the body solve the problem of lifting a suitcase of unknown weight, of pushing open a swing door whose springs offer unknown resistance, or of perpetually operating against gravity which puts a continual limitation on movement? We understand very little about the details, but the answers seem to lie in the functioning of the cerebellum which lies attached to the hind-brain just above the spinal cord. The cerebellum receives a steady supply of all the information which could conceivably be of importance in muscle movement. Using this information it modifies the instructions sent out by the motor cortex so that the power of muscular activity is continually adjusted to cope with varying opposing forces.

The cerebellum is different from most other parts of the brain in that each side of it receives information from and controls the same side of the body, not the opposite side as in the case of the cerebral cortex. The left hand side of the motor cortex controls the right hand side of the body but the left hand side of the cerebellum controls the left hand side of the body. Damage to the left side of the cerebellum leads to weakness on the left side of the body. As a result, when a person with such damage walks, he veers to the left because the right leg takes normal steps while the left leg takes small ones. In contrast the patient tends to look to his right: this is because the muscles on the right side of the neck are working normally while those on the left are weak and so the head is pulled to the right. The reverse situations will occur if the right side of the cerebellum is damaged.

Normally movements are smooth because the cerebellum continually adjusts the strength of muscle contraction to deal with the resistance which the muscles meet. These adjustments are carried out entirely subconsciously and depend on information from the muscles and joints and many other receptors. If the cerebellum is damaged, the smooth adjustments no longer take place and movements become jerky as the person has to adjust the strength of muscle contraction consciously. In a person with cerebellar damage, the muscles show no tremor (shaking) at rest, but as soon as movement begins a jerkiness or tremor appears. Because this occurs only on movement it is sometimes known as 'Intention tremor'. A good test which usually clearly shows up intention tremor is to ask the patient to touch the tip of his nose with his fore-finger. As you can demonstrate on yourself a normal person can do this quickly and smoothly either with eyes shut or with eyes open. Someone with cerebellar damage may just succeed when his eyes are open, but the movement will be slow and jerky and will become jerkier and jerkier as the nose is approached and precise adjustments are required. With eyes shut a person with cerebellar damage usually completely fails to touch his nose: this is because in the absence of the cerebellum he relies entirely on his eyes to tell him about the progress of a movement.

Like all other muscles, those of the larynx, mouth and face are also controlled by the cerebellum. In people with cerebellar damage the movements of these muscles cannot be regulated precisely and speech becomes slurred (dysarthria).

One part of the cerebellum, known as the flocculonodular lobe, is particularly concerned with the maintenance of the upright posture. If it is damaged the person finds it impossible to stand upright even though while he is sitting the muscle movements of arms and head are normal. There is a tumour of small children known as a medulloblastoma which characteristically starts in the flocculonodular lobe. The first indication of its presence is usually difficulty in standing upright even though muscle movements when lying in bed are more or less normal.

Muscle Spindles and Posture

Muscle spindles are also important in the maintenance of normal posture. As we saw earlier when discussing the 'knee jerk' when a muscle is stretched, the muscle spindles in that muscle discharge impulses which travel to the spinal cord. There they synapse with the moto-neurons going back to the same muscle that is stretched. A burst of impulses is therefore fired along these moto-neurons and the muscle contracts. This reflex is important in helping you to stand upright. If when standing you sway forwards, the muscles in the back and on the back of the legs are slightly stretched. This acti-vates the muscle spindles and as a result a reflex occurs which makes those muscles contract and bring you back to the upright position. If you sway forwards or to one side, similar events occur which bring you upright again. The muscle spindles and the reflex they initiate are there-fore important in the maintenance of posture.

Basal Ganglia

These are a group of nerve centres in the fore-brain near the thalamus. Their function is poorly understood but it is certain that like the cerebellum they have an important role to play in the maintenance of posture and the regu-lation of movement. They are important in medicine because it seems that the basal ganglia are at fault in the very common condition known as Parkinson's disease. This disease is characterised by the following three major features:

1. Muscle stiffness. The limbs are stiff and difficult to move and in the face this appears as a 'woodenness' of expression.

2. Difficulty in maintaining the upright position with a tendency to fall frequently.

3. Tremor which, in contrast to that of cerebellar disease, is present at rest even when the patient is not attempting to carry out any movement.

The condition may sometimes be considerably improved by surgical destruction of some of the basal ganglia which appear to be malfunctioning. It has been suggested that it may be due to a deficiency of a chemical in the brain known as L-DOPA and treatment with this drug has recently given encouraging results.

DAMAGE TO THE NERVOUS SYSTEM

There are innumerable forms of damage to the nervous system but here I shall discuss only three common ones where a knowlege of physiology is particularly helpful in understanding what happens.

PERIPHERAL NERVE DAMAGE

When a peripheral nerve is cut, stimuli which fall on the area of skin which that nerve supplies can no longer be felt. The muscles supplied by that nerve are paralysed and can no longer contract in either voluntary or reflex movements. They become totally flabby (flaccid). The muscle may undergo spontaneous irregular contractions as a result of impulses being fired off in the irritated cut end of the nerve but unless regeneration of the nerve occurs the muscle will eventually atrophy and all the contractile tissue will be replaced by fat.

Loss of nerve supply to the skin causes the skin to become shiny and to lose elasticity: it is taut and easily damaged. The mechanism of this skin change is unknown but it may be partly caused by lack of secretion of the glands in the skin.

If the two ends of a cut nerve can be sewn together some recovery may occur. The central end of the nerve remains alive because it is still connected to the cell bodies in or near the spinal cord. The peripheral part dies but the tubes which contained the nerve axons remain. If regeneration takes place, new axons growing from the central end may travel down the empty tubes and eventu-ally reach the nerve endings and re-establish function. However, regeneration is a slow process and complete recovery is extremely unlikely. This is obvious from the fact that a nerve the thickness of a pin may contain 100,000 fibres: for effective function each fibre must travel from the spinal cord to a particular part of the skin, to a particular muscle or to a particular gland without interruption. It is obvious that if the two ends of a cut nerve are stitched together only a tiny proportion of the fibres will succeed in making the right connections.

Regeneration of nerves is often such a prolonged process that by the time even a few fibres reach a muscle, the muscle has atrophied. In order to prevent this, physio-therapists may stimulate a muscle by giving direct elec-tric shocks to it. This does not completely prevent atrophy but the regular contractions slow down the process considerably.

SPINAL CORD SECTION

Complete transection of the spinal cord produces total paralysis of voluntary movement and total absence of conscious sensation at points innervated from below the level of the injury. No nerve regeneration at all occurs in the CNS and if the cord has truly been severed the damage is permanent. Such a condition where the lower part of the body is paralysed while the upper part is normal is known as paraplegia.

Immediately after damage to the spinal cord, the muscles cannot be moved voluntarily, nor do they move in response to reflexes and are completely paralysed and flaccid. This situation of total unresponsiveness of muscles is known as spinal shock. It lasts only a few minutes in the frog, an hour or so in cats and dogs and two to six weeks in man. During this period no reflexes can be demonstrated. The bladder fills but does not empty and must be drained by means of a catheter. Similarly faeces must be removed manually from the rectum because defaecation does not take place. Even-tually however the spinal shock passes off and in most cases the reflexes listed return.

1. *Babinski response.* This is a sign of damage to the pyramidal tract and has already been discussed.

2. *Flexion reflex.* This is often very troublesome as it may become hypersensitive. Normally it is set off only by painful stimuli but in paraplegics it may be initiated by very minor things such as breadcrumbs in the bed or ingrowing toe nails. Although the patient cannot feel pain, violent flexion reflexes may throw him around the bed.

3. *Micturition.* When the bladder fills to a certain critical level it empties automatically as in an infant. The patient may be able to achieve some control over micturition, as in some paraplegics the reflex may be initiated by scratching or pressing on the anterior abdominal wall, thus enabling the patient to micturate at a convenient time.

4. *Defaecation.* This too becomes automatic. A few patients may be able to initiate it by pressing hard on the abdomen to increase the intra-abdominal pressure.

5. *Ejaculation of sperm.* Mechanical stimulation of the penis may produce erection and ejaculation even though the patient can feel nothing. If the sperm is collected it can then be used for artificial insemination of the wife so that a paraplegic patient may have children.

6. *Stretch reflex* (tendon jerk). In the later stages of recovery this too may become overexcitable. As a result the limbs become very stiff and resistant to bending and the tendon jerks become greatly exaggerated. This condition is known as spasticity.

CEREBROVASCULAR ACCIDENT (CVA) OR STROKE

A stroke is caused by damage to the brain resulting from blockage of an artery by a clot or from the rupture of an artery with resulting haemorrhage. The damaged area is most commonly in the region of the tracts coming down from the motor cortex and going up to the sensory cortex. As a result, the patient loses voluntary movement and conscious sensation on the opposite side of the body. This situation, where say the right half of the body is damaged while the left half is normal, is known as hemiplegia in contrast to the paraplegia which results from spinal section. The main features of hemiplegia apart from loss of voluntary movement and conscious sensation, are spasticity and a Babinski response on the affected side. Speech is very commonly affected. If the nondominant cerebral hemisphere is affected, the defect is likely to be dysarthria because the motor control of the muscles is abnormal but the understanding and use of words is normal. If the dominant hemisphere is affected, however, the patient may have trouble in understanding and selecting words in addition to the motor defect: this more serious situation is known as aphasia.

If the stroke occurs lower down in the brain in the region of the pons, the normal functioning of the reticular formation may be abolished. As a result the patient may lapse into a deep and permanent state of unconsciousness.

PAIN

The sensation of pain depends on the activation of fine nerve endings which can be found throughout the body. These fire off impulses whenever tissue damage has occurred or is about to occur. The usual stimuli which cause pain are extremes of heat and cold and mechanical stimulation and certain types of chemical damage such as erosion by acid. The impulses are carried to the spinal cord by sensory nerves which synapse with the fibres of the spinothalamic tract. The tract immediately crosses to the opposite side of the spinal cord and travels up to the thalamus.

All sensations have what may be called factual and emotional components. A touch on your hand activates the same receptors no matter who is doing the touching: that is the factual component. But your reaction to the touch depends very much on who is doing the touching: that is the emotional component and it depends on your assessment of the significance for your life of a factual event. Pain has a very strong emotional component. Not only are we aware of the fact of a painful stimulus but we are also aware that it may mean discomfort, illness or even death. In some cases the emotional component of pain may be over-ruled by even stronger emotional circumstances and there are many examples of this in war and in sport. A soldier in battle or a footballer in an important match may be unaware of quite severe injuries: the man realises that they have happened only when all the excitement is over. This demonstrates that the intensity of the pain sensation depends a great deal on the circumstances. Severe pain in the leg due to cramp is annoying but not emotionally disturbing: pain of a similar intensity in the chest may throw the patient into an acute anxiety state because he knows that it may mean a heart attack and death.

It therefore follows that in relieving pain two quite different things must be done:

1. The patient must be reassured that he is in good hands and that the cause of the pain is being dealt with calmly, quickly and effectively. This reassurance may be aided by giving sedative drugs which dull the patient's awareness of the significance of the pain.

2. Specific pain-relieving drugs may be given. These fall into four main categories:

a. Those which are temperature-lowering (antipyretic) as well as pain-relieving (analgesic). Aspirin (acetyl salicylic acid) is the most used member of this group although there are many others. They relieve pain partly by acting directly on pain receptors and partly by acting on the brain.

b. Those which are sleep-inducing (narcotic) as well as analgesic. Morphine and heroin come into this group. They are effective because they both allay fear and also dull the factual component of the pain probably by acting on the parts of the brain which deal with pain.

c. Local anaesthetics. These act directly on the peripheral nerves which carry information about pain. They block nerve conduction and are mainly used in dentistry and for the stitching of minor wounds.

d. Anaesthetics which relieve pain by producing total unconsciousness. These are used for surgical operations.

SURGICAL RELIEF OF PAIN

Sometimes, particularly in patients with intractable cancer, pain may be so severe and persistent that drugs become ineffective. In these patients surgical procedures on the nervous system may be considered. There are three main types of operations:

1. Destruction of the spinothalamic tract by cutting the anterolateral part of the spinal cord where the tract runs. This operation is known as antero-lateral cordotomy.

2. Destruction of the part of the thalamus associated with pain. This is usually done by passing long needle electrodes into the thalamus and then passing strong electric currents through them in order to destroy brain tissue in the vicinity of the tip.

3. Prefrontal leucotomy. This severs the connections between the frontal part of the cerebral cortex and the rest of the CNS. The operation does not interfere with the factual part of pain sensation but it does alter its emotional component. The patient still feels the pain but is no longer unduly worried by it. The operation is sometimes done in other conditions, notably severe depression but it is now tending to fall out of use.

REFERRED AND PROJECTED PAIN

Two other types of pain are important for the nurse and doctor. They are referred pain and projected pain. With referred pain, pain which really originates in some deep structure is felt somewhere on the skin surface: it is said to be 'referred' to the skin. There are many examples of this, but some important ones are:

1. Pain arising in the heart may be felt in the neck, the jaw, the shoulders (particularly the left) and down the inside of the left arm.

2. Pain arising from injury to or inflammation of the diaphragm (the sheet of muscle dividing chest from abdomen) may be referred to the tip of the shoulder.

3. In the early stages of appendicitis, pain arising from the inflamed appendix may be referred to the umbilicus.

4. Pain arising from the hip joint may be referred to the knee.

Referred pain seems to occur because there are more pain fibres in peripheral nerves than there are in the spinothalamic tract. Thus several peripheral nerves, some from

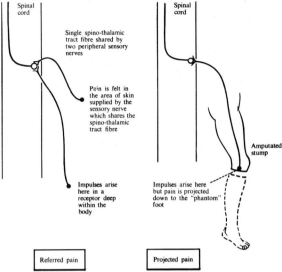

Figure 4.45 Referred and projected pain.

deep structures and some from the skin, must share the same spinothalamic fibre. During life, skin pain receptors are activated much more frequently than deep ones so that the thalamus comes to associate pain in a particular spinothalamic fibre with injury to the skin. When pain does arise in a deep structure it may be the first time in fifty or sixty years that the deep pain receptors have been activated: for example pain receptors in the heart are not activated until angina or a coronary thrombosis occurs. As a result the thalamus tends to refer the pain to the skin.

Projected pain is the pain which arises whenever the pain pathway is cut or damaged somewhere on its route from the pain receptor to the thalamus. A good example is that after amputation of a leg, the patient may complain that his foot is hurting (phantom limb pain). This is a very real sensation: the pain is genuinely felt and the patient must not be dismissed as a fraud. The pain occurs because the cut ends of nerves in the limb stump continue to fire impulses for some time. The brain 'knows' that impulses in those fibres normally come from pain receptors in the foot and so the pain is 'projected' to the foot and actually felt there. Another good example of projected pain is sciatica, when pressure on the sciatic nerve as it leaves the vertebral column at the bottom of the back fires off impulses in pain fibres in the nerve. The brain 'knows' that impulses in those fibres normally come from receptors in the back of the leg and so the pain is projected down the back of the leg.

AUTONOMIC NERVOUS SYSTEM AND THE ADRENAL MEDULLA

All the motor nerves going to skeletal muscles are capable of being brought under voluntary control. A single nerve fibre goes right from the spinal cord to the muscle it supplies. In contrast, with a few exceptions, the motor nerves which control the behaviour of smooth muscle and of glands are not usually under voluntary control. These latter nerves make up the autonomic nervous system. They also differ from skeletal muscle nerves (sometimes called somatic nerves) in that there are usually at least two nerve fibres for carrying the impulses between the spinal cord and the muscle or gland.

The autonomic nervous system itself is divided into two main divisions, sympathetic and parasympathetic.

1. *Parasympathetic*

This arises from the brain and from the sacral region of the spinal cord. The main parasympathetic nerve arising in the brain is the vagus. The most important actions of the parasympathetic system are on the eye, the heart rate, the gut, the bladder and the reproductive organs. The nerve fibre which leaves the CNS is always very long and carries the parasympathetic impulses right to the organ where they will act. In the organ there is a synapse and the impulses are transmitted to a second short fibre which then goes to a muscle or to a gland cell. The chemical transmitter released by both first and second fibres is acetyl choline.

2. *Sympathetic*

This arises from the thoracic and lumbar regions of the spinal cord. The first nerve fibre is short and travels to one of the sympathetic ganglia (a ganglion is a collection of synapses) which lies in two chains, one on either side of the vertebral column. There are some additional ganglia in the mid-line, the most important of which is the coeliac plexus which lies in front of the aorta where the coeliac artery leaves the aorta. The chemical transmitter in the ganglia which carries the impulses from the first to the second neurons in the pathway is again acetyl choline. The sympathetic fibres which leave the ganglia (post-ganglionic fibres) then go to smooth muscle and gland cells where they almost all act by releasing noradrenaline. The main exceptions to this rule are the sympathetic fibres to the sweat glands which release acetyl choline.

THE ADRENAL MEDULLA

An endocrine gland which is very closely associated with the sympathetic system is the adrenal medulla, the inner part of the adrenal gland. The outer part, the adrenal cortex, has quite different functions (see next chapter). The adrenal medulla is supplied by pre-ganglionic sympathetic fibres which come straight from the spinal cord without synapsing. They release acetyl choline which stimulates the adrenal medulla to secrete its two hormones, adrenaline and noradrenaline.

ACTIONS OF THE AUTONOMIC SYSTEM

The main actions of the autonomic nervous system and the adrenal medulla are summarised in table 4.2. One problem is that adrenaline seems to have two quite different groups of actions on smooth muscle. It makes many smooth muscle fibres, such as those in skin blood vessels, contract while it makes others, such as those in the bronchi and in muscle blood vessels, relax. Noradrenaline

Table 4.2 The actions of the autonomic nervous system and of circulating adrenaline. Circulating adrenaline can activate both alpha and beta receptors but noradrenaline released from sympathetic nerves activates only alpha receptors. ACh indicates acetyl choline (2.1 EP).

Structure	Receptor type	Sympathetic	Parasympathetic
Heart S-A node	Beta	Increases heart rate	—
	ACh	—	Slows the heart
Heart muscle	Beta	Increases contractility	—
Muscle arterioles	Alpha and beta	Constriction (noradrenaline) Dilatation (adrenaline)	
Skin, gut and kidney arterioles	Alpha	Constriction	
Bronchial muscle	Beta	Relaxation	—
	ACh	—	Contraction
Bronchial glands	ACh	—	Secretion
Main gut muscle	Beta	Relaxation	—
	ACh	—	Contraction
Salivary glands	Alpha	Mucus secretion	
	ACh	—	Watery secretion
Gastric glands	ACh	—	Secretion
Pancreas	ACh	—	Secretion
Sweat glands	ACh	Secretion	
Bladder wall	Beta	Relaxation	—
	ACh	—	Contraction
Sphincter of bladder	Alpha	Contraction	—
	ACh	—	Relaxation
Male sex organs	?	Ejaculation	Erection

in contrast seems to be purely excitatory, making all types of smooth muscle contract. a synthetic substance, isoprenaline (isoproterenol) makes all smooth muscle fibres relax.

It is now apparent that muscle fibres contain two different types of keys or receptors on to which adrenaline, noradrenaline, isoprenaline and related substances can become attached. These receptors have been called alpha (α) and beta (β). In general, α receptors cause muscle contraction while β receptors cause muscle relaxation. The structure of the noradrenaline molecule is such that it can occupy only α receptors and can only cause excitation. The isoprenaline molecule can occupy only β receptors and can only cause relaxation. Adrenaline, however, can occupy both α and β receptors and the overall effect which it has depends on whether there are more α receptors or more β receptors on that muscle. In lung smooth muscle, the β receptors predominate and so adrenaline causes relaxation. In skin arterioles α receptors predominate and so adrenaline causes contraction.

There is one important exception to the general rule that β receptors are inhibitory and α receptors excitatory.

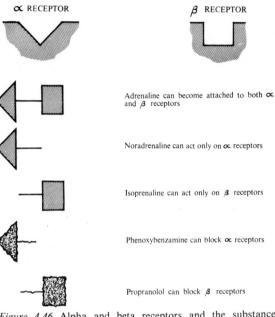

ȣ RECEPTOR β RECEPTOR

Adrenaline can become attached to both ȣ and β receptors

Noradrenaline can act only on ȣ receptors

Isoprenaline can act only on β receptors

Phenoxybenzamine can block ȣ receptors

Propranolol can block β receptors

Figure 4.46 Alpha and beta receptors and the substances which act on them.

This is that the β receptors in the heart which are involved in the control of heart rate are excitatory. Adrenaline and isoprenaline both therefore increase the rate of beating of the heart.

5 The Endocrine System

The hormonal or endocrine system represents one of the two great control systems in the body. The nervous system is the other. The nervous system exerts its control by actually sending nerve fibres to the various organs, but the hormonal system exerts its control at a distance. The endocrine glands release their secretions (hormones or 'chemical messengers') directly into the blood stream. The blood then carries the hormones to every part of the body. The nervous and endocrine systems, although apparently so different, are not entirely separate. They come together in the connection between the hypothalamus and the pituitary gland.

THE THYROID GLAND
The thyroid is perhaps the best known of all the endocrine glands, partly because of its position at the front of the neck and partly because disorders of it are relatively common. It consists of two lobes lying on either side of the trachea (windpipe) and joined by a narrow strip called the isthmus. It receives a very copious blood supply. Surgically it is important to remember that the four parathyroid glands lie embedded on its posterior surface and that the recurrent laryngeal nerve which supplies many of the muscles of the larynx lies very close. Both parathyroids and the nerve may be easily damaged at operation.

IODINE AND THYROID HORMONE
Iodine is essential for the manufacture of thyroid hormone. It is found primarily in sea food and in the sea, and the quantities in food and drinking water diminish on moving inland. It is well absorbed from the gut and as it circulates in the blood it is trapped and concentrated in the cells of the thyroid gland. There it is attached to a protein containing the amino acid tyrosine. This iodinated protein is stored in a material called colloid. The colloid itself is found in the centre of spheres of thyroid cells known as follicles. When the hormone is required, it is split off from the colloid and released into the blood. Therefore in an actively secreting gland there is little col-

loid and the follicles are small. In a resting gland the colloid accumulates and the follicles become large.

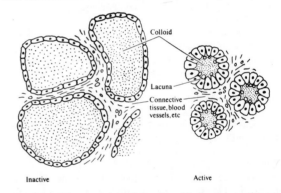

Figure 5.1 The structure of the thyroid gland in active and inactive states.

The thyroid hormone is secreted in two forms, thyroxine which contains four iodine atoms and tri-iodo-thyronine which contains three: both types contain two tyrosine molecules attached together. Over 90% of the hormone is in the form of thyroxine and almost all this is carried in the blood closely bound to the plasma proteins. Therefore if the amount of iodine which is bound to protein is estimated (protein-bound iodine or PBI), this gives a good indication of the level of hormone in the blood. The normal range for the PBI is 3–8ug/100 ml.

ACTIONS OF THYROID HORMONE
Very many different actions have been described but the most important ones are:

1. Thyroid hormone (TH) can alter the basal metabolic rate. An exess of TH produces a hot, hyperexcitable individual while a lack of TH produces a cold, dull one.

2. TH is essential for normal growth and in this it co-operates with growth hormone and insulin. However,

while growth hormone deficiency leads to a person of small size with normal intellectual and emotional development, TH deficiency leads to severe retardation of growth and of both emotional and mental development. Children who have suffered from thyroid deficiency since infancy are known as cretins.

3. Even in adults normal blood levels of TH are required for the normal functioning of the nervous system. If TH is deficient the person becomes mentally sluggish but is subject to sudden rages. Excess TH leads to overexcitability and nervousness.

4. TH stimulates glucose absorption from the gut so that very high blood levels of glucose may occur after a carbohydrate meal. This glucose may appear in the urine but it does not indicate diabetes as the blood level usually comes down rapidly again.

5. TH lowers plasma cholesterol levels . If TH is deficient, plasma cholesterol levels may become very high and this is a useful test of thyroid deficiency.

6. TH is required for the conversion of carotene to vitamin A. In the absence of TH carotene accumulates and may give a yellowish tinge to the skin. In severe cases signs of vitamin deficiency may appear.

7. TH increases the effectiveness of adrenaline and of the sympathetic nervous system. These actions are particularly evident in the case of the heart which may beat very rapidly in thyrotoxicosis: it may even start fibrillating.

CONTROL OF THE THYROID GLAND
The body clearly requires a method of controlling the amount of TH secreted by the thyroid. It does this with the aid of the pituitary gland lying beneath the hypothalamus at the base of the brain.

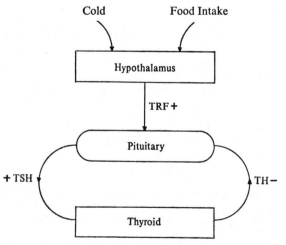

Figure 5.2 The control of the output of thyroid hormone. Plus signs indicate stimulation, minus signs indicate suppression.

The pituitary exerts an influence on the body out of all proportion to its size. The anterior part of the pituitary releases into the blood a substance known either as thyrotrophic hormone, thyroid stimulating hormone or more usually simply as TSH. An increase in the amount of TSH in the blood stimulates the thyroid to start secreting more hormone. A decrease in the amount of TSH leads to a decrease in the amount of TH released by the thyroid. In turn, the pituitary seems able to measure the amount of TH in the blood and controls its output of TSH accordingly. If the TH level rises, the pituitary releases less TSH so bringing the TH level back to normal. A fall in TH on the other hand stimulates the output of more TSH.

But even the pituitary is not completely independent. It receives blood via a special system of vessels from the hypothalamus. The hypothalamus can secrete into this blood another substance known as TSH-releasing factor or TRF. When the output of TRF rises, so do the outputs of TSH and TH. The secretion of TRF if increased under two main circumstances:

1. Prolonged exposure to cold. The effect is to raise the metabolic rate and so increase the amount of heat produced by the body.

2. Prolonged increase in food intake. Again the metabolic rate tends to increase in order to burn the food up more effectively. Unfortunately this mechanism seems to work much better in some people than in others. Some can therefore eat without getting fat because or their high metabolic rate, while others with a low metabolic rate must watch their weight very carefully.

CLINICAL ASPECTS
Disorders of the thyroid gland can have four main origins. There may be a lack of iodine in the diet or the hypothalamus, anterior pituitary or thyroid itself may be functioning abnormally. These defects may show themselves in three main ways:

1. Signs of excessive output of thyroid hormone (hyperthyroidism or thyrotoxicosis).

2. Signs of insufficient TH output (hypothyroidism, cretinism in childhood, myxoedema in adults).

3. A swelling in the neck (goitre) without obvious signs of hypo- or hyperthyroidism.

GOITRE
This is the term used for a visible swelling of the thyroid. If it is associated with an excessive output of TH it is said to be a toxic goitre. If the output of TH is normal or low, the goitre is said to be non-toxic.

Most goitres are non-toxic and are due to iodine deficiency in the diet. Levels of TH may not be quite sufficient and in an effort to raise them to normal, the pituitary pours out TSH. As a result, the gland is enlarged and if the deficiency is prolonged it may reach an enormous size. In some individuals the output of TH may normally be just sufficient but under the impact of some additional strain such as pregnancy or puberty when additional amounts of TH are required, the gland may again enlarge.

Other goitres may be caused by tumours of the thyroid gland or by an abnormal pituitary which pours out large amounts of TSH even though blood TH levels may be above normal. Such goitres are often toxic.

HYPERTHYROIDISM

This may be caused by overactivity of the hypothalamus, pituitary or thyroid itself and in most cases the precise reason for the disturbance cannot be identified. In yet another group of patients the blood contains a substance of unknown origin (possibly from the liver) called long-acting thyroid stimulator (LATS): this has a similar action to TSH in that it stimulates the thyroid to pour out TH. The main features of hyperthyroidism are:

1. The patient has a high metabolic rate, sweats a lot and dislikes hot weather.

2. The patient is excessively active, both physically and mentally.

3. There is a rapid pulse rate, possibly with atrial fibrillation.

4. The eyeballs protrude. This is partly due to excessive sympathetic nerve activity which pulls on the muscles of the upper eyelid exposing more of the eyeball than usual: this applies in all types of hyperthyroidism. However, in some cases the protrusion is much worse because the disease is usually of hypothalamic or anterior pituitary origin and the fat deposition may occur because of the presence of some abnormal anterior pituitary secretion.

There are several different ways of treating thyrotoxicosis. There are a number of antithyroid drugs available (such as carbimazole) which interfere with the manufacture of TH and so reduce its level in the body. Unfortunately the patient usually has to take the drugs for life. Secondly, radioactive iodine can be used. Quite a large dose is employed and the radioactive material is concentrated in the thyroid gland where the radioactivity destroys some of the cells. This is a simple and usually permanent treatment. However, it is difficult to estimate the correct dose of radioactivity and a disturbingly high proportion of the patients treated become myxoedematous later on. The third method of treatment is partial thyroidectomy in which a large part of the gland is surgically removed. Before operation the thyrotoxicosis is brought under control by the use of antithyroid drugs. The blood supply to the gland is also reduced by giving very large doses of iodine (much larger than those normally required in the diet). How the iodine acts is a mystery but it greatly reduces bleeding and makes the operation much easier.

HYPOTHYROIDISM

There are two main forms, cretinism in children and myxoedema in adults. Cretinism may be due either to maternal iodine deficiency or to a congenital defect in the thyroid gland which renders it incapable of manufacturing TH. The main characteristics of cretinism are small stature, mental deficiency, an infantile face (because of a failure of bone development), a coarse skin, a slow pulse and sluggish gut. The only treatment is thyroid hormone and a full cure is achieved only if it is started early in infancy.

In adults hypothyroidism is usually termed myxoedema although this refers to only one of its features, the thickening and puffiness of the skin. The skin in these patients in addition to being puffy is dry, waxy and cool. Oedema of the vocal cords leads to a striking deepening and huskiness of the voice. The patient is sluggish both physically and mentally and may be given to sudden rages. The pulse rate is slow and the cholesterol level in the blood is high, a useful diagnostic test. Because of the high cholesterol level the patients seem to be unusually susceptible to fatty changes in the atreries and coronary thrombosis.

THE ADRENAL CORTEX

The adrenal gland consists of two separate parts of quite different functions. The inner medulla is really part of the sympathetic nervous system and is discussed in chapter 4. The outer section of the gland, the adrenal cortex, is very much part of the endocrine system. It produces three different groups of steroid hormones, the mineralocorticoids, glucocorticoids and the androgens.

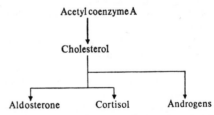

Figure 5.3 The three main types of adrenal cortical hormone are all manufactured from acetyl CoA and cholesterol.

MINERALOCORTICOIDS

The most important of the mineralocorticoids is known as aldosterone. The hormones are so-called because they are primarily concerned with the mineral content of the body, in other words with the inorganic materials. Their main concern is with the body content of sodium ions. They stimulate the kidney to remove sodium from the urine and so to retain it in the body. If there is a large amount of sodium in the body, the output of aldosterone is low, so allowing the excess sodium to escape in the urine. If there is too little sodium in the body, the output of aldosterone is high in order to prevent any unnecessary loss in the urine.

The control of the output of aldosterone is as yet poorly understood. It seems to be relatively little affected by ACTH which is important in controlling the other adrenal hormones (see later in this chapter). A low sodium concentration in the blood going to the adrenal can raise aldosterone output while a high sodium concentration can reduce the secretion of the hormone: this mechanism makes obvious sense but the changes in concentration required to cause the effect seem large in comparison with those which occur naturally. At present, the most important mechanism of controlling aldosterone output seems to depend on renin, an enzyme which is produced by the kidneys. The output of renin from the

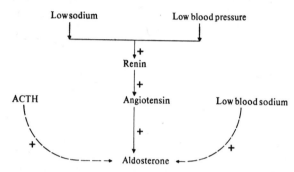

Figure 5.4 The ways in which aldosterone output may be controlled.

kidneys rises if the total sodium content of the body is low: it also rises if the blood pressure is low, as a low blood pressure may be an indication of sodium deficiency. The renin then acts on a plasma protein and breaks off from this protein a substance called angiotensin which is a powerful constrictor of blood vessels. In addition, angiotensin has an action on the adrenal cortex where it stimulates the output of aldosterone. This helps to retain sodium in the body and returns the body sodium content and the blood pressure to normal. In contrast, rises in body sodium content and in arterial pressure can suppress the output of aldosterone by the same renin-angiotensin mechanism, the output of renin being greatly reduced.

GLUCOCORTICOIDS
Glucocorticoids are so-called because they have important actions on carbohydrate metabolism, although that is not their only function. The most important member of the group is cortisol, often known as hydrocortisone. Cortisone is not secreted naturally but is converted to cortisol in the body. Recently some synthetic steroids have been made (e.g. prednisone and prednisolone) which are much more powerful glucocorticoids than cortisol itself. The main actions of the glucocorticoids are:

1. They are essential for the normal excretion of water by the kidney. In their absence the body can get rid of water only slowly.

2. They are essential for the maintenance of a normal blood pressure although how they act is unknown.

3. They are required for the manufacture of red blood cells.

4. They are essential to enable the body to respond to any type of stress whether it be pregnancy, a surgical operation, an uncomfortable climate or a bad emotional experience. In all these circumstances the plasma cortisol levels are raised although, again, its precise mode of action is unknown.

In higher concentrations than are normally found in the body, the glucocorticoids have other actions as well:

1. They block the inflammatory response. The inflammatory response consists of an increased blood flow and migration of white blood cells to a damaged area of the body, coupled with a laying down of fibrous tissue. In most cases, such as in the healing of wounds or in fighting bacterial infections, this response is beneficial to the body, but in some cases such as the inflammation of the joints in rheumatoid arthritis or the inflammation of the heart valves in rheumatic fever, the process is harmful: it serves no useful purpose as there are not usually any bacteria in the joints or on the heart valves. Steroids such as cortisol may therefore be used to suppress this harmful type of inflammation. Inevitably, at the same time, they reduce the body's ability to cope with infections, especially tuberculosis. Those who are treated with steroids for long periods of time should therefore have regular chest X-rays.

2. They interfere with the manufacture of proteins and so muscles become weak. The protein collagen, so essential for the strength of blood vessels and of bone is also weak. Bleeding therefore tends to occur and skeletal defects may appear.

3. They cause loss of calcium and phosphate from the kidney. The blood level of calcium is maintained by removal of calcium from bone and this further aggravates the weakness of bone. This weakness is particularly apparent in the vertebral column and in severe cases one or more vertebrae may collapse.

4. They raise the blood glucose level and cause large amounts of fat to be deposited, especially round the shoulders and over the abdomen.

Control of Glucocorticoid Output
The control of cortisol secretion is similar in many ways to the control of TH secretion. The anterior pituitary secretes a substance known as adrenal corticotrophic hormone or ACTH. ACTH stimulates the adrenal to secrete cortisol. In turn, the ACTH output depends on the secretion of ACTH releasing factor (usually called CRF) from the hypothalamus. Finally the output of CRF

depends on the blood level of cortisol. If plasma cortisol levels are too high, the output of CRF falls and with it the outputs of ACTH and cortisol, so returning the plasma

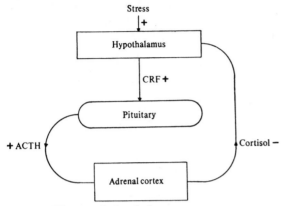

Figure 5.5 The control of cortisol output.

cortisol level to normal. If cortisol levels are too low, the output of CRF and ACTH rises, so bringing cortisol levels up again. During periods of stress, the brain directly increases the output of CRF so making more cortisol available to cope with the emergency.

ANDROGENS

The word androgen means a substance which has an action like that of the male sex hormones. However, despite this name, the adrenal androgens seem to be important in both sexes although relatively little is known about their function. Their main actions appear to be:

1. Stimulation of pubic and axillary hair growth at puberty. Hair grows at puberty in females even in the absence of ovaries (Turner's syndrome) and in males even in the absence of testes.

2. The stimulation of grease production by the skin. Some think that acne is caused by excessive production of adrenal androgens.

3. The development of muscles.

4. The development of sexual desire in both males and females.

CLINICAL ASPECTS OF ADRENAL FUNCTION

A number of clinical conditions are associated with disorders of the adrenal glands. The main ones are discussed in this section.

Addison's Disease

This is due to a failure of the adrenal to secrete its hormones. It is usually due to destruction of the gland, often by tuberculosis but frequently for unknown reasons. The most important consequences are a massive loss of

sodium in the urine coupled with a very low blood pressure. In chronic cases, anaemia, low blood glucose, failure to respond to stress and pigmentation of the skin are also apparent. The skin pigmentation develops because the pituitary pours out ACTH in an effort to raise the output of adrenal hormones to normal. ACTH has a skin-darkening action in large doses and results in a peculiar bronzed appearance, giving a false impression of good health. The disease can be fairly satisfactorily treated by means of the various hormone preparations available.

Conn's Syndrome

This is due to a tumour which appears to pour out excessive amounts of mineralocorticoids alone. As a result there is salt retention and high blood pressure. Because of the way in which the kidney works, the retention of large amounts of salt in the body is often associated with a loss of potassium in the urine. Thus the blood level of potassium is often low, resulting in cramps and muscular weakness. The condition can be cured by removal of the tumour.

Cushing's Syndrome

This is primarily due to an excess secretion of glucocorticoids, although mineralocorticoids and androgens are often also secreted in above normal amounts. It may occur because of an adrenal tumour, a pituitary tumour or a cancer (often a lung cancer) which for no apparent reason suddenly starts pouring out large amounts of ACTH. Lung cancers are the commonest type to behave in this way. The main features of Cushing's syndrome are:

1. A high blood glucose, coupled with obesity. The limbs tend to be relatively thin, contrasting oddly with the grossly fat face and trunk.

2. Suppression of the inflammatory response with a resulting susceptibility to infections.

3. Easy bruising and bleeding with weak muscles.

4. Backache because of loss of calcium from the vertebral column.

5. High blood pressure.

6. A high red cell count.

7. The development of hair on the face and other parts of the body. This is naturally more obvious in females.

Androgenic excess

This may occur in early infancy or even in utero when it is usually due to a congenital defect in one of the enzymes manufacturing aldosterone or cortisol. As a result these hormones cannot be manufactured and the adrenal cortex devotes all its energies to making androgens. In female children, the external genitalia become masculinised with growth of the clitoris. Male genitalia hypertrophy. In both sexes hair development and muscle growth occur as in adults. In later childhood, similar changes may be brought about by androgen secreting tumours. In adult males androgen-secreting tumours often go unnoticed until a late stage but in females they cause pronounced masculinisaton early on.

CALCIUM AND THE PARATHYROIDS

Calcium is important in the body in many different ways. The main ones are:

1. It is essential for the normal activity of many enzymes.

2. It is important in maintaining the stability of cell membranes.

3. It is required for muscular contraction and for the normal functioning of the nervous system.

4. It is essential for blood clotting.

5. It is essential for the manufacture of strong bones and teeth.

The total concentration of calcium in the plasma is 9–11 mg/100 ml. It exists in two main forms, as free calcium ions and as calcium bound to plasma protein. The two types can interact as follows:

Calcium ions + protein → Calcium-protein complex

Only the free calcium ions are physiologically active. The balance between free and protein-bound calcium depends largely on the pH of the blood. If for any reason the blood becomes more alkaline, more calcium is bound by the protein and the concentration of free calcium ions may fall below normal. The main situations when the blood may become more alkaline are:

1. When acid is lost because of repeated vomiting of gastric juice.

2. When exessive amounts of the potentially acid gas, carbon dioxide, are lost from the body because of overbreathing. This is most likely to occur in hysteria and during childbirth when the mother often overbreathes.

The concentration of free calcium ions will also fall if the total plasma calcium falls. This occurs when the parathyroid glands fail to function normally. Whenever the concentration of free calcium ions falls, whatever its cause, the condition of tetany results. The symptoms are due to overactivity of the nervous system. Nerve impulses fire off spontaneously in both motor and sensory nerves. The abnormal sensory impulses cause tingling sensations often known as 'paraesthesiae' while the motor impulses cause involuntary muscle twitches. These twitches are usually most apparent in the muscles of the inner side of the hand and fore-arm supplied by the ulnar nerve. As a result the fourth and fifth fingers curl up

giving the 'main d'accoucheur' so called because it is the position of the hand when doing a vaginal examination. The facial muscles also often twitch.

When due to overbreathing, tetany is very easily cured by simply putting a paper bag over the mouth and nose for a few minutes. The carbon dioxide breathed out is then breathed straight back in again. This makes the blood less alkaline and the link between calcium and protein is broken. The concentration of free calcium ions returns to its normal level and the tetany disappears. Tetany due to a low calcium concentration may be rapidly but temporarily relieved by injecting a solution of a calcium salt (e.g. calcium gluconate) intravenously.

BONE

Bone contains three main types of cell and three main types of extracellular material. The basic structure of bone consists of a thick interlacing web of fibres of the protein collagen. These fibres are glued together by a special type of carbohydrate known as a mucopolysaccharide. Finally they are made hard and durable by the deposition on their surfaces of mineral substances, primarily calcium and phosphate. The three main types of cell are the osteocytes, osteoblasts and osteoclasts. The osteocytes predominate in bone which is stable and is neither being built up nor destroyed: they seem in an unknown way to maintain the health of the bone. The osteoblasts predominate where new bone is being laid down while the osteoclasts are large multinucleated cells which seem to be involved in the breakdown of old bone.

There are two main types of bone, the long bones and the flat bones. The flat bones grow because their inner surfaces are eroded while new bone is laid down on their outer surface. The long bones grow primarily at the epiphysial regions. Each long bone has at least at one end, and often at both, a spearate piece of bone known as the epiphysis. New bone formation occurs mainly in the region between the shaft of the long bone and its epiphysis. This region is rich in cartilage rather than bone. After puberty when growth ceases, the epiphyses become firmly united to the shafts by proper bone.

CALCIUM ABSORPTION

The main sources of calcium are dairy foods. Like iron, calcium is poorly absorbed but there are two factors which help the process. Vitamin D is the most important of these. A normal secretion of acid gastric juice is helpful but not essential: this is because in an alkaline medium calcium-phosphate combinations tend to be very insoluble while they are more soluble as the pH falls. Thus patients who have little natural acid secretion or whose stomachs have been removed at operation are in danger of becoming calcium deficient.

Three main factors can hinder the normal absorption of calcium:

1. The presence of large amounts of unabsorbable fat in the gut. The fat forms an insoluble complex with the calcium and stops the calcium being absorbed as well.

This situation may occur when fat is poorly digested because of lack of bile or of pancreatic juice. It may also occur if the wall of the intestine is damaged as in sprue or coeliac disease.

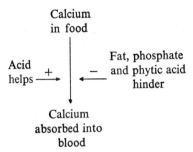

Figure 5.6 Factors affecting the absorption of calcium from the gut.

2. The presence of large amounts of phosphate in the food which also forms an insoluble combination with calcium.

3. The presence of excess phytic acid in food which yet again forms an insoluble complex. Phytic acid is found primarily in wheat flour and in order to combat its action, extra calcium is added to the bread in many countries.

VITAMIN D
This has two main actions on calcium metabolism. In the gut it is essential for calcium absorption and in its absence enough calcium cannot be taken into the body. Once in the body, it is required for the normal manufacture and calcification of new growing bone, especially at the epiphyses. In its absence normal bone growth cannot occur, giving the disease of rickets. The long bones and the pelvis in particular may become permanently deformed.

PARATHYROID HORMONE
The parathyroid glands are essential for the maintenance of normal plasma calcium levels. They are four tiny structures normally embedded in the back of the thyroid. Not

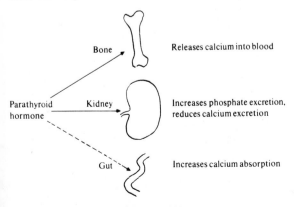

Figure 5.7 The actions of parathyroid hormone.

unusually, however, one or more may be found elsewhere in the neck or in the thorax. The glands secrete a hormone known as parathormone which has three main actions:

1. It releases calcium from bone into the blood.

2. It reduces the excretion of calcium by the kidneys.

These first and second actions tend to raise the plasma calcium level.

3. It causes the excretion of phosphate by the kidneys. This lowers the plasma phosphate level and is important because calcium and phosphate together tend to precipitate out of solution as an insoluble complex. This can damage many tissues and arterial walls in particular. This is obviously a highly undesirable event and it can be prevented if the phosphate concentration is lowered as the calcium concentration in the plasma rises. This is the action of parathormone.

Parathormone may have a minor action in increasing calcium absorption from the gut but this is not thought to be very important. The output of parathormone depends

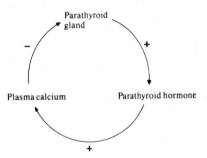

Figure 5.8 The control of blood calcium by parathyroid hormone.

on the level of calcium in the blood. If the plasma calcium falls, the output of parathormone increases and raises the calcium back to normal. If the plasma calcium rises, the secretion of parathormone is suppressed so allowing calcium levels to fall back to normal.

CALCITONIN
This is a relatively recently discovered hormone which may be made by both thyroid and parathyroid glands. Its action is to lower plasma calcium by causing the deposition of calcium in bone. Its importance in medicine has not yet been worked out.

PARATHYROID DISORDERS
Underactivity of the parathyroids (hypoparathyroidism) may be caused by atrophy of the glands for unknown reasons or by accidental removal of the glands during partial thyroidectomy. The plasma calcium level falls and tetany results. The tetany may be temporarily relieved by

an intravenous injection of calcium gluconate. In the long term the condition must be treated by a very high calcium diet and large doses of vitamin D.

Hyperparathyroidism is much rarer and is usually due to a tumour. It is commoner in women. Calcium is removed from the bones causing weakness: it is then deposited elsewhere in the body, particularly in arterial walls, in the kidneys and in the urine. For unknown reasons peptic ulcers are particularly common. The only treatment is removal of the tumour.

REGULATION OF BLOOD GLUCOSE

Normally the concentration of blood glucose is within the range of 60–90 mg/100 ml of plasma. The maintenance of a reasonably constant level is one of the most important functions of the hormonal system. This is so that at all times all organs of the body can obtain adequate supplies of food from which they can obtain energy. The constancy is particularly important for the cells of the brain which seem able to oxidise glucose only and which cannot make use of fat. Almost all the cells of the body apart from those of the liver and brain present some barrier to the entry of glucose from the blood. Many of the hormones act by altering the effectiveness of this barrier so changing the ease with which the glucose can enter the cells. Cells of the liver and brain are freely permeable to glucose at all times.

The main problem of blood glucose regulation results from the fact that meals are taken relatively infrequently. Immediately after a meal large amounts of glucose enter the blood and the plasma concentration tends to rise. But within a short time the absorption of the meal is completed and the blood glucose level tends to become very low. The aim of the regulating system is to iron out these wild swings, to lower the blood glucose level just after a meal and to raise it between meals.

The first line of defence against these swings is provided by the liver which receives all the portal vein blood from the gut. When glucose is being absorbed very rapidly just after a meal, much of it is taken up by the liver and stored in the form of glycogen: the blood leaving the liver via the hepatic veins contains less glucose than the blood entering the liver via the portal vein. In contrast, when absorption has been completed and no glucose is entering the portal blood from the gut, the liver breaks down some of its glycogen and releases it into the blood as glucose. In this way a sharp fall in blood glucose can be avoided.

The liver is aided in this smoothing-out action by the activities of a number of hormones.

INSULIN

This is a protein hormone released by tiny groups of cells in the pancreas known as the islets of Langerhans. The islets are quite distinct in structure from the main part of the pancreas which secretes digestive juices along the pancreatic duct. The most important action of insulin is to lower the blood sugar. It does this in two ways:

1. It stimulates the liver to manufacture glycogen from the glucose which reaches the liver in the blood.

2. It stimulates most of the cells in the body, particularly those in muscles, to take up glucose quickly. In the absence of insulin many cells in the body (but not those of the brain and liver) become almost impermeable to glucose and cannot take it up from the blood. They must therefore use fat instead for their energy supply. In skeletal muscles and in the heart much of the glucose taken into the cells is converted into a glycogen store.

The other main actions of insulin are to stimulate the manufacture of proteins from the amino acids absorbed after a meal and to stimulate the formation of triglycerides in adipose tissue from the fatty acids in the blood. Insulin therefore tends to increase the availability of carbohydrate inside cells and to reduce the availability of fat.

The output of insulin is controlled by the level of blood glucose itself. When the blood glucose level rises, as after

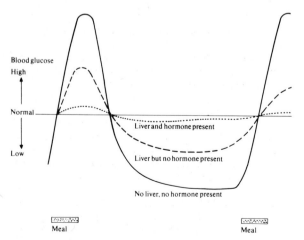

Figure 5.9 The blood glucose level is normally kept nearly constant. In the absence of the liver and of hormonal control the wild swings shown in the diagram would occur.

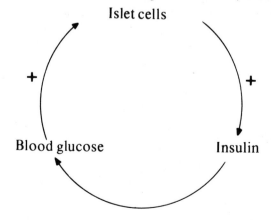

Figure 5.10 The control of insulin output.

a meal, the insulin output also rises, so tending to bring the blood glucose back to normal. When the blood glucose falls, the rate of insulin secretion also falls: this helps to stop the cells using glucose and so the blood glucose level tends to rise again.

GROWTH HORMONE

This is secreted by the anterior pituitary gland and in some ways its actions are the reverse of the actions of insulin. The main ones are:

1. It stops most of the cells in the body (like those of muscle) taking glucose from the blood. This halts the fall in blood glucose. It has no effect on the brain cells and so there is plenty of glucose available for the nervous system.

2. It causes the breakdown of fat to free fatty acids. These enter the blood and most cells, muscle in particular, use them as an alternative energy supply.

3. It continues to stimulate protein synthesis and so does not interfere with this important aspect of metabolism.

The output of growth hormone is also controlled by the blood glucose level but the effects are precisely opposite to those which control insulin output. The output of

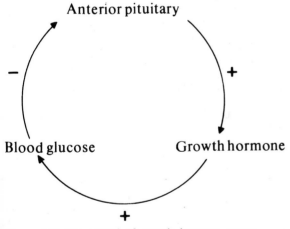

Anterior pituitary

Blood glucose   Growth hormone

Figure 5.11 The control of growth hormone output.

growth hormone is increased when the blood level falls and decreased when the blood glucose level rises. So just after a meal when blood glucose concentration tends to be high, insulin levels are also high and growth hormone output is low. The high insulin concentration stimulates the cells to use glucose. But when the meal has been absorbed and blood glucose levels fall, the insulin level also falls while growth hormone output increases. Thus there is a barrier to the entry of glucose into cells but an increase in the availability of fats. The body thus tends to use fat instead of glucose: at the same time the glucose concentration in the blood is maintained for the use of the

brain cells which can use nothing else. The day to day control of glucose levels and the balance between the uses of fat and of carbohydrates depend primarily on a close co-operation between insulin and growth hormone.

ADRENALINE (EPINEPHRINE)

This is not routinely important in the control of blood glucose but it can raise the concentration rapidly in an emergency when the brain is in danger of not getting enough energy. It acts in two ways:

1. It stimulates the liver to break down glycogen to glucose quickly. It also stimulates glycogen breakdown in muscle so increasing the availability of glucose to the muscle cells.

2. It reduces the rate at which cells like those of muscle remove glucose from the blood: they must rely more on their own stores.

These actions produce a rapid rise in blood glucose so that cells in the nervous system can receive an adequate supply. The output of adrenaline is stimulated by a low blood glucose concentration and the release is accompanied of course by other actions of adrenaline and of the sympathetic nervous system. A hypoglycaemic (low blood glucose) attack is therefore accompanied by sweating and by a thumping, rapid heart beat. There is usually a feeling of faintness because of the lack of energy supply to the nerve cells.

OTHER HORMONES

Two other hormones can effect glucose metabolism but their precise significance under normal conditions is uncertain.

1. *Glucagon* is also produced by the islets of Langerhans but it has precisely the opposite actions to insulin. It raises blood sugar and it even stimulates the breakdown of protein to give glucose so that it has been suggested that it may be of use in starvation.

2. *Cortisol.* High levels of cortisol produce a high blood glucose as in Cushing's syndrome. Low cortisol levels as in Addison's disease lead to a low blood glucose level. Despite this the precise role of cortisol in glucose metabolism is uncertain.

DIABETES MELLITUS

The word diabetes used to be used for any condition in which the output of urine was excessive. Two varieties were known; in one the urine was sweet (mellitus) and in the other it was tasteless (insipidus). Diabetes insipidus is much rarer than diabetes mellitus and is due to lack of antidiuretic hormone from the pituitary (see next section). When the word diabetes is used alone it usually refers to diabetes mellitus.

Diabetes mellitus is a state in which the insulin levels appear to be too low to cope with the patient's intake of carbohydrate. As a result five important things tend to occur:

1. The blood sugar level is much higher than normal.

2. If the blood sugar level rises above 180 mg/100 ml, the kidneys cannot hold glucose back from the urine and so glucose appears in the urine. This is abnormal and the glucose can be picked up by several different tests (Benedict's, Fehling's, 'Clinitest' and 'Clinistix').

3. When glucose appears in the urine, excess water must also be excreted to carry away the glucose in solution. The patient therefore complains about producing large amounts of urine.

4. The large amounts of water lost make the patient very thirsty and he complains of drinking a lot.

5. Because much of the glucose taken into the body is lost, more calories must be supplied by excessive eating.

In early diabetes all these features may not be present and there may be no sugar in the urine if it is tested at random. The diabetic state may then be uncovered by means of a glucose tolerance test. In this, a patient is

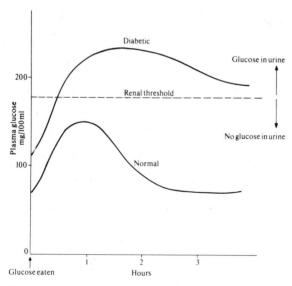

Figure 5.12 The glucose tolerance test.

given a large amount of glucose either by mouth or intravenously. Urine and blood samples are then taken at half-hourly intervals for 2½–3 hours. Normally the blood glucose concentration does not rise above about 150 mg/100 ml and then rapidly returns to normal: no glucose appears in the urine. In diabetes the blood glucose becomes very high, stays high for a long period, and glucose overflows into the urine.

In addition to the features mentioned above, diabetes may cause many complications most of which are not fully understood. The main ones are:

1. Deposition of fat in arteries, leading to heart disease and the blocking of major arteries, especially in the legs.

2. Damage to the kidneys ending in renal failure.

3. Damage to the retina ending in blindness.

4. Repeated infections, especially of the skin and urine. This seems to be because bacteria find the glucose-rich body fluids an ideal place to grow.

5. Ketosis and diabetic coma. This is perhaps the best-known complication and its mechanism is now fairly well understood. It begins because there is not enough insulin to enable most of the cells in the body to use carbohydrate: as a result there is a switch to the oxidation of fat instead. Because of the lack of carbohydrate breakdown there may not be enough oxaloacetic acid to combine with acetyl coenzyme A to allow for complete fat breakdown by the citric acid cycle. The accumulation of acetyl coenzyme A leads to the formation of the ketone bodies (acetone, acetoacetic acid and beta-hydroxybutyric acid): these make the blood acid and in addition are poisonous to the brain. In consequence, the patient becomes comatose. There are usually three vital things wrong with a patient in diabetic coma:
    a. The high levels of toxic ketone bodies.
    b. The acidity of the blood.
    c. The high levels of blood glucose which usually cause massive fluid loss in the urine so that the patient is severely dehydrated.
The treatment is therefore directed to correcting these three things:
    a. Insulin is given in large amounts to switch the body metabolism to carbohydrate use and to burn up the ketone bodies.
    b. Intravenous fluids are given to make up the deficiency.
    c. Some of the fluid is usually given in the form of alkaline bicarbonate, lactate or citrate solutions which help to counteract the acidity of the blood.

Types of Diabetes

Diabetics fall into two quite distinct categories. The so-called 'juvenile' type genuinely do have a lack of insulin. They are usually very thin because of the large losses of glucose and can be treated only by injections of insulin. Insulin cannot be given by mouth because it is digested in the gut.

In the much commoner 'adult' or 'mature' type, the main problem does not seem to be an insulin lack. In fact insulin levels are often higher than in normal individuals. Such diabetics tend to be obese. There are several possible explanations for this mature type, none of which is fully accepted.

1. There may be a very high carbohydrate intake which exceeds the capacity of insulin to cope with it.

2. The cells may be resistant to the action of insulin so that much higher levels of the hormone may be required to produce a normal effect.

3. There may be present in the blood insulin antagonists which prevent insulin exerting its normal action. This certainly seems able to account for the type of diabetes seen in Cushing's syndrome when the excess cortisol levels persistently push up the plasma glucose levels. An increased output of growth hormone which also tends to raise blood glucose may explain some other cases. For example, it is well-known that a baby whose birth weight is over 10 lb almost always indicates that the mother is a diabetic or will develop diabetes sooner or later. This could be due to an excessive maternal output of growth hormone.

Since mature diabetes is not due to an actual absolute lack of insulin it may be possible to treat it without insulin injections. In fact there are four main ways of tackling it:

1. By cutting down the intake of carbohydrate in the food (especially of sucrose) so reducing the need for insulin.

2. By using oral drugs like tolbutamide which stimulate the islets of Langerhans to pour out more insulin.

3. By using oral drugs like phenformin which have an insulin-like action and which stimulate the uptake of glucose by cells.

4. By insulin injections. In practice most mature diabetics can be satisfactorily treated by some combination of the first three measures, so avoiding the need for injections of insulin.

THE PITUITARY AND HYPOTHALAMUS

The pituitary is the single most important endocrine gland in the whole body. This is because in addition to producing hormones which act in their own right, it also manufactures hormones (so-called trophic hormones) which control the behaviour of the thyroid, the adrenal cortex, the ovaries and the testes. In turn the pituitary itself is controlled by the part of the brain known as the hypothalamus and so in the end the brain is intimately involved in controlling the behaviour of the endocrine system.

The pituitary has two quite distinct parts, anterior and posterior, sometimes called the adenohypophysis and the neurohypophysis. During fetal life, the posterior pituitary develops as a downgrowth from the hypothalamus. The anterior pituitary develops as an upgrowth from the roof of the mouth. The posterior pituitary receives a very rich nerve supply from the hypothalamus. The anterior pituitary seems to have no nerve supply but instead is intimately connected to the hypothalamus by a complex

system of blood vessels known as the pituitary portal system. The hypothalamus releases chemicals known as releasing factors (RF) into these vessels. They are then

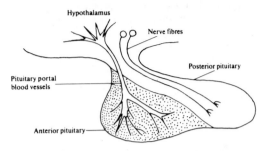

Figure 5.13 The structure, blood supply and nerve supply of the pituitary gland.

carried by the blood down the pituitary stalk to the anterior pituitary where they stimulate the output of the anterior pituitary hormones.

THE ANTERIOR PITUITARY

This produces two hormones which act in their own right and four trophic hormones which act by controlling the behaviour of their endocrine glands.

Growth Hormone

This is essential for the normal growth of a child. If it is deficient, a dwarf results: if it is present in excess the outcome is a giant. Apart from their size, these individuals are normal human beings. This is in marked contrast to the retardation of mental and emotional development which occurs in thyroid deficiency. If excessive output of growth hormone occurs after the epiphyses of the long bones have fused and growth in height has ceased, growth hormone can then have much effect only on the soft tissues and on the bones of the hands, feet and face. Thus the nose, hands, feet and soft tissues such as the tongue may become enormous in size. This condition is known as acromegaly.

Apart from its action in stimulating growth, growth hormone is also important in the regulation of blood sugar. It is essential for the switch from predominantly carbohydrate to predominantly fat oxidation which occurs once a meal has been completely absorbed. In acromegaly when the output of growth hormone is persistently high, diabetes frequently occurs: the growth hormone keeps the blood glucose level high and in an attempt to secrete enough insulin to keep the glucose level down, the islets of Langerhans appear to become exhausted.

Prolactin

This is essential for the development of the breasts during pregnancy and for the secretion of milk once birth has taken place. The output of prolactin is stimulated by suckling. The sensory nerves in the nipples send nerve impulses up to the hypothalamus which then stimulates the output of prolactin from the anterior pituitary. If suckling stops, prolactin secretion falls to very low levels and within a short time the breasts cease to secrete milk. It has recently been shown that prolactin may help to regulate sodium, potassium and water excretion.

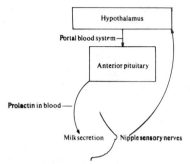

Figure 5.14 The stimulation of milk secretion by prolactin.

Adrenal Corticotrophic Hormone (ACTH)

This is essential for the normal control of the output of cortisol from the adrenal cortex. It may also help in the stimulation of the output of aldosterone and the androgens but in these cases it probably does not act alone. In turn, the output of ACTH is controlled by the secretion of corticotrophin releasing factor (CRF) by the hypothalamus into the pituitary portal blood vessels.

Thyroid Stimulating Hormone (TSH)

This stimulates the thyroid gland to secrete thyroid hormone. The output of TSH itself is in turn controlled by the level of thyroid hormone in the blood and also by the level of the TSH releasing factor (TRF) secreted by the hypothalamus.

Follicle Stimulating Hormone (FSH)

This is the hormone which is essential for the development of the egg-bearing follicles in the ovary. It is also required for the manufacture of sperm by the testis in the male. Even though it is called FSH after its action in the female, it is now known that the FSH in males is identical to that in females: it is the ovaries and testes which differ in their response to it.

Luteinizing Hormone (LH)

This is the hormone which stimulates the output of hormones from the ovary and from the testis. The ovary secretes primarily oestrogens and progesterone while the testis secretes testosterone. Again LH is identical in males and females: in males it is sometimes called interstitial cell stimulating hormone or ICSH. The actions of these sex hormones are further discussed in chapter 10.

Sometimes, since prolactin, LH and FSH are all concerned with sexual function, they are lumped together under the name of gonadotrophic hormones.

THE POSTERIOR PITUITARY

This produces two important hormones, antidiuretic hormone (ADH or vasopressin) and oxytocin. Both these hormones are made by nerve cells in the hypothalamus. They then travel down the nerve fibres which pass along the pituitary stalk to the posterior pituitary. The hormones are actually released into the blood in the posterior pituitary gland.

Oxytocin

This is important in three ways. During delivery of a baby, sensory receptors in the wall of the uterus and in particular in the cervix are activated. They send nerve impulses up to the hypothalamus which then stimulates the output of oxytocin from the posterior pituitary. The oxytocin makes the uterus contract more vigorously, so helping to push the baby out.

Secondly, oxytocin is essential for the ejection of milk from the breasts. Prolactin stimulates the manufacture of

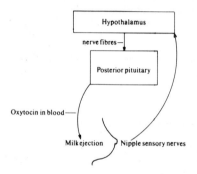

Figure 5.15 The stimulation of milk ejection by oxytocin.

milk but it cannot expel the milk from the breasts out of the nipples. The milk simply accumulates in the ducts in the glands and makes the breasts engorged and painful. However, the ducts are surrounded by cells with a muscular type of action known as myoepithelial cells. When the baby sucks the nipple, sensory receptors in the nipple are activated. They send nerve impulses to the hypothalamus which then stimulates oxytocin secretion. The oxytocin travels in the blood to the breasts and makes the myoepithelial cells contract, thus forcing milk out of the breast. This is why when a baby is suckling or a cow is being milked, it takes a minute or two before the milk flows freely.

Lastly, oxytocin appears to be important during sexual intercourse. It is probably partly responsible for the female sensation of orgasm which is associated with contractions of the uterus.

Oxytocin is also present in males but its function, if any, is uncertain.

Antidiuretic Hormone

This hormone acts on the kidney to reduce the output of water in the urine. If the body water content is low, the plasma becomes slightly concentrated. This change in plasma concentration is detected by the hypothalamus which increases the output of ADH. The ADH travels in the blood to the kidney and reduces the amount of water lost in the urine, making the urine smaller in volume and more concentrated. On the other hand if an excess of fluid is drunk, this dilutes the blood a little. This change too is detected by the hypothalamus which reduces the amount of ADH secreted, so allowing the kidney to excrete a large volume of dilute urine. If ADH is absent because of damage to the hypothalamus or posterior pituitary, vast amounts of dilute urine are secreted, a condition known as diabetes insipidus. Because of the tremendous loss of water the patient is perpetually thirsty.

6 The Body Fluids and Blood

In a lean adult male, about 60% of the body weight consists of water. Fatty tissue contains very little water and so as the amount of fat in the body rises, so the proportion of body weight made up of water, falls. Females are almost always rather fatter than males and so in a normal female 55% of the body weight consists of water. In an extremely obese person, irrespective of sex, the water content of the body may fall to 40% of the body weight.

INTRACELLULAR AND EXTRACELLULAR WATER

The total body water is divided into two major compartments, the water within cells (intracellular water) and that outside cells (extracellular water). About 55% of the water is intracellular and about 45% is extracellular. The extracellular water is further split up as follows:

1. Inside the blood vessels in the plasma (7·5% of total body water),

2. Interstitial fluid which bathes cells and lymph (20% of total body water),

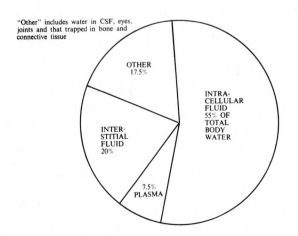

Figure 6.1 The main body fluid compartments.

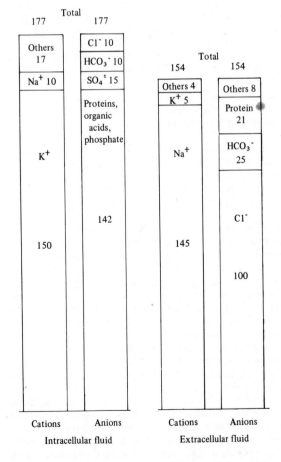

Figure 6.2 The compositions of intracellular and extracellular fluid. All figures are in mEq/L.

3. Other compartments (17·5%). These include the cerebrospinal fluid, fluid in the eyes and joints, water bound to connective tissue and water bound to bone.

The ionic compositions of the plasma and interstitial fluid which are in free communication with one another are similar but are very different from the composition of the intracellular fluid. The compositions of the intracellular and interstitial fluids are compared in fig. 6.2. The main points to note are the high concentration of sodium outside the cell and the high concentration of potassium inside the cell and the high concentration of negatively charged organic substances in the cell which cannot cross the cell membrane.

INTERCHANGE BETWEEN CELLS AND INTERSTITIAL FLUID

If a membrane separates two solutions, several factors will govern the movement of water and dissolved particles across the membrane. The most important are:

1. *Membrane permeability.* If a membrane is permeable to a substance the substance will be able to move freely from one side to the other under the influence of electrical and concentration gradients described in 2. and 3. If a membrane is not permeable, the substance will be able to move from one side to the other.

2. *Concentration gradients.* All substances tend to move from areas where their concentration is higher to ones where their concentration is lower.

3. *Electrical gradients.* Like charges repel and unlike charges attract. Positively charged particles will therefore tend to move away from positively charged areas towards areas of negative charge. Negatively charged particles will behave in the opposite way.

4. *Osmotic forces.* These come into operation when a membrane which is permeable to water separates two solutions containing different concentrations of a dissolved substance which cannot cross the membrane. When considering osmotic forces only particles which cannot cross the particular membrane concerned are relevant. The chemical nature of the particles, on the other hand, is quite irrelevant. When calculating osmotic forces what matters is the total concentration on one side of a membrane of particles of any chemical composition which cannot cross the membrane, as compared with the total concentration of similar particles on the other side. When a difference in such particle concentration exists, then water will move from the side of the membrane where the particle concentration is higher to the side where it is lower. Under appropriate conditions this water movement may be prevented by the application of hydrostatic pressure. The osmotic pressure is defined as the pressure which would have to be applied to block osmotic water movements in any given situation.

Now consider the situation between the intracellular fluid and the interstitial fluid. The cell membrane is definitely impermeable to the large negatively charged organic ions within the cell and therefore these must exert an osmotic force tending to pull water into the cell. Since living cells do not normally swell and burst because water steadily moves into them, it is clear that there must be some other force which opposes the osmotic pressure of the organic anions. This force is also osmotic and depends on the fact that although dead cell membranes are freely permeable to sodium, living cell membranes are not. There seems to be a mechanism associated with the cell membrane which, with the expenditure of metabolic energy, persistently pumps sodium ions out of the cell. This means that the living cell membrane is effectively impermeable to sodium and therefore that the sodium ions in the extracellular fluid exert an effective osmotic pressure which balances the osmotic pressure exerted by the internal organic anions and enables the cell to remain in fluid equilibrium. The high internal concentration of potassium inside cells is partly a consequence of the fact that if a lot of positively charged sodium ions are outside the cell, then because of electrical forces other positive ions will be repelled and inevitably tend to move into cells. It is also partly due to a potassium 'pump' which is much weaker and less effective than the sodium one but which does tend to move potassium from interstitial fluid into cells.

The sodium pump operates only in living cells. If a cell dies, its membrane becomes freely permeable to sodium and sodium moves into it: in exchange potassium moves out. Furthermore the organic anions are now the only effective osmotically active particles and therefore water moves into the cell until it either bursts or its expansion is physically limited by contact with other cells. One of the outstanding features of dead cells examined by the pathologist is that they become grossly swollen as compared to normal living cells.

THE PLASMA

The main difference between the composition of the plasma (blood without the red and white cells and platelets) and the interstitial fluid, is the high concentration of protein in the former. Normal plasma contains 6–7 g of protein per 100 ml. Two main types of protein are present, the relatively low molecular weight albumins (MW about 60–70,000) and the higher molecular weight globulins (MW up to 900,000 but mainly in the region of 90,000–180,000). Most of the globulins are antibodies (immunoglobulins) formed in the course of immune responses.

The main function of the albumin seems to be to exert an osmotic pressure in order to maintain the fluid balance across the capillary wall (see next section). It may also combine with and so help to transport around the body bilirubin, fatty acids and some hormones.

By weight, the globulins make up in the region of 40% of the total amount of plasma protein. However, because globulin molecules are much larger than albumin ones, the globulins contribute less than 20% to the total osmotic pressure exerted by the plasma proteins. Their prime importance is in immunity but some also have a role in the transport of metals, such as iron and copper, and of some hormones, such as thyroid hormone.

There are many other protein substances found in plasma but their concentrations are low in comparison with albumins and globulins. They include:

1. *Various clotting factors* such as prothrombin (see later this chapter).

2. *Plasminogen.* This is an inactive substance which under appropriate conditions can be converted to active plasmin which can dissolve blood clots. As yet the details of the mechanism are obscure but much research is being done at the moment because of the importance of clots in heart disease and strokes.

3. *Cholinesterase.* This destroys acetyl choline. It also happens to destroy succinyl choline, a neuromuscular blocking agent which is widely used for producing brief paralysis during surgical operations. In one out of every two thousand or so people the enzyme is congenitally absent. In normal life this seems to cause no difficulty but such patients undergoing surgery may be paralysed for unusually long periods after the normal dose of succinyl choline.

CAPILLARY FUNCTION

It is particularly important to understand the inter-relationships between the plasma and the interstitial fluid. Most of the blood vessels are effectively impermeable to the passage of water and dissolved solids across their walls. It is only the tiniest ones, the capillaries with walls a single cell thick, which allow fluid and dissolved material to pass from plasma to interstitial fluid and vice versa.

Since the blood volume normally remains approximately constant, it is clear that the amount of fluid lost from the plasma into the interstitial fluid and the amount entering the plasma from the interstitial fluid must be evenly balanced. How is this balance maintained? Five factors must be considered, the last two being relatively unimportant:

1. *The permeability of the capillary wall.* The wall is normally freely permeable to water, to ions such as those of sodium, chloride, and bicarbonate and to small organic molecules such as those of glucose, amino acids and free fatty acids. In contrast it is almost impermeable to the plasma proteins although the smaller molecules can escape in tiny quantities. In effect, all the constituents of the blood apart from the cells and the protein can freely pass out of the capillaries into the interstitial fluid.

2. *Protein osmotic pressure.* Because virtually no protein can cross the capillary wall, there is very little protein in the interstitial fluid. As a result the osmotic pressure

exerted by the plasma proteins tends to draw water from the interstitial fluid into the blood.

3. *Blood pressure.* The pressure of the blood in the capillaries, about 30 mm Hg at the arterial end and 15 mm Hg at the verous end, is much higher than that of the interstitial fluid. This force therefore tends to force fluid from the blood out into the interstitial fluid.

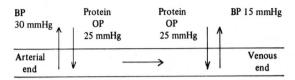

Figure 6.3 The function of a capillary.

4. *Interstitial fluid protein osmotic pressure.* This force is normally negligible but may become important if the protein content of the fluid rises because of loss from the plasma, as can occur in burns and local inflammation.

5. *Interstitial fluid pressure.* This force is normally negligible in comparison with the blood pressure and in most cases is actually negative. However, if abnormal amounts of interstitial fluid accumulate as in oedema or inflammation, the interstitial fluid pressure may rise and become a significant factor.

Under normal circumstances there is an approximate balance between the blood pressure pushing fluid out of the capillaries and the plasma protein osmotic pressure drawing fluid back in. The blood pressure tends to be higher than the osmotic pressure at the arterial end of a capillary and so fluid tends to move out there. At the venous end of the capillary the blood pressure has fallen and the osmotic pressure of the proteins, which is unchanged or may have risen a little because of water loss, is usually higher than the blood pressure. This results in fluid being drawn back into the blood.

THE LYMPH

Usually the blood pressure pushes out of the capillaries very slightly more fluid than the osmotic pressure of the proteins draws back in. This means that fluid tends slowly to move out of the blood into the interstitial fluid. If a disaster is not to occur, this fluid must somehow be returned to the blood and this is where the lymphatic system comes in.

The lymphatic system is a blind ended system of tubes ramifying to all parts of the body. Its finest branches, the lymphatic capillaries, are like the blood capillaries in consisting of a single layer of cells. Unlike the blood capillaries they have very permeable walls and any protein which is in the interstitial fluid can freely enter the lymph. Particles of dirt, red and white blood cells, bacteria and cancer cells can also gain entry to the lymphatic capillaries. The lymphatic capillaries collect not only fluid but all sorts of foreign or waste material which gains entry to the interstitial fluid. In a way the lymphatic vessels therefore act as a drainage system. It would clearly be unfortunate if all this material were simply poured into the blood without some form of purification. This is one of the functions of the lymph nodes. The nodes form a complex system of filters which remove or destroy most of the noxious material before it can enter the blood.

In principle, the nodes function by greatly widening the lymph channels so that the flow rate slows right down. The wide channels in the nodes are known as sinuses. Their effect is similar to the effect of a rapid river suddenly widening and slowing down as it enters a lake. All the sediment which the lymph has been carrying, such as cancer cells or bacteria, settles out. Much of this sediment is then engulfed by special cells which line the walls of the sinuses and are known as reticulo-endothelial or phagocytic cells. This is why, when a cancer or an infection develops in any part of the body, it tends to be temporarily stopped at the lymph nodes. In the case of infection, the nodes which receive the lymph draining from the infected area become swollen and painful. In the case of cancer, the growth develops in the nodes and may be prevented for some time from gaining entry to the blood.

The lymph which leaves the nodes is thus largely purified. The lymph channels come together and grow larger and eventually the lymphatic vessels pour lymph into the venous system at several sites. The most important one is via the thoracic duct which empties into the left jugular vein at the side of the neck: this receives the lymph from most of the lower part of the body.

OEDEMA
Oedema is the word used to describe the excessive accumulation of fluid in the tissues. It has three main causes, each clearly related to one of the factors which govern the passage of fluid across the capillary wall.

1. Capillary permeability. If the capillary wall is damaged and becomes freely permeable to protein, then plasma proteins can no longer effectively exert an osmotic pressure. The blood pressure can thus act unopposed and fluid is lost rapidly from the capillaries: the fluid which is lost is effectively protein-rich plasma. Capillary damage happens particularly during burns and local infections.

2. Raised pressure of capillary blood. If the pressure of blood in the capillaries rises, then fluid will move out into the tissues more easily. Since capillary permeability is unchanged, the fluid lost has little or no protein in it. A rise in arterial pressure is usually not transmitted to the capillaries because the arterioles constrict to prevent such transmission. In contrast, if venous pressure rises as it does in heart failure or over expansion of blood volume, the capillaries have no protection and the raised venous pressure is thus transmitted directly to them. This is why oedema is one of the outstanding features of heart failure. The reverse movement of fluid may occur after blood loss when capillary pressure may become very low.

3. Plasma protein concentration. If the concentration of plasma proteins falls the osmotic pressure they exert also falls. The force drawing fluid into the blood is therefore reduced and fluid escapes from the capillaries more easily. This can occur in starvation and may help to account for the gross oedema sometimes seen in that condition.

THE BLOOD
The blood is the central means of transport within the body and it puts every organ in contact with every other organ. It has two main components, the plasma and the so-called 'formed elements'. The formed elements include:

1. Red cells which contain the pigment haemoglobin which carries oxygen from the lungs to the tissues and carbon dioxide in the reverse direction.

2. White cells which are primarily important in combating the invasion of foreign organisms and materials. They are much less numerous than the red cells.

3. Platelets which are pale non-nucleated cell fragments manufactured by the cells known as megakaryocytes in the bone marrow. They are essential for the normal control of bleeding from a wound.

BLOOD VOLUME, HAEMATOCRIT AND RED CELL COUNT
The total blood volume in a normal adult is in the region of 4–5 litres, depending on size. The simplest way of roughly estimating the amount of red cells in the blood is to take a small sample of blood, treat it with something to prevent clotting and to centrifuge it in a specially calibrated tube. The red cells settle out at the bottom of the tube. On top of them is an exceedingly thin layer of white cells which is sometimes known as 'the buffy coat'. Above the white cells is the clear, straw-coloured plasma.

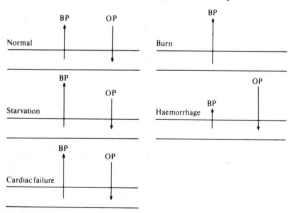

Figure 6.4 Imbalances between blood pressure and protein osmotic pressure seen in some abnormal situations.

If the tube is calibrated it is easy to read off the percentage of red cells and plasma by volume: the volume of the white cells is well under 1% of that of the whole blood. Usually about 45% of the blood volume is made up of red cells and this is known as the haematocrit value.

The red cell count is the number of red blood cells in one cubic millimetre of blood. The count is usually in the region of 4·5 to 6 million/mm³. The count can be made manually using a microscope and a special slide with a counting chamber but increasingly automatic counting machines are coming into use.

ERYTHROCYTE SEDIMENTATION RATE (ESR)

The red blood cells are often known as erythrocytes. Because they are heavier than the plasma, if blood is treated with something to prevent it clotting and then simply left, the red cells will gradually settle out. This has been made the basis of the simple standard test of the erythrocyte sedimentation rate or ESR. The blood is drawn up into a narrow tube either 10 cm or 20 cm long depending on the precise technique being used. The tube is then left in the upright position for one hour and then the amount of clear plasma left above the red cells is noted. With a 20 cm tube, in normal young individuals the rate of fall is about 5 mm/hour. The ESR increases slowly with age, but rates over 15 mm/hour are probably always abnormal.

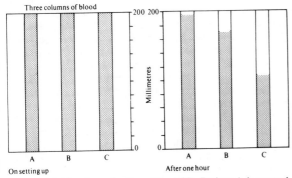

Figure 6.5 The determination of the ESR. Patient A is normal, B has a slightly raised ESR and C has a greatly raised ESR.

In diseases the composition of the plasma proteins, and especially of the globulin component, changes as the body responds to the illness. For reasons which are still uncertain, these changes make the red cells more sticky and when they are left to stand they tend to come together on clumps. These clumps fall more rapidly than single red cells and so the sedimentation rate increases. In many diseases and particularly in cancer, coronary thrombosis and inflammatory conditions, the ESR is always high. If and as the patient gets better the ESR gradually returns to normal.

The ESR is especially useful in two situations:

1. Many patients who consult doctors appear to be vaguely ill but it is impossible to diagnose their trouble with certainty. If the ESR is abnormally high, then there is certainly something wrong and the patient should be studied until it is found. A normal ESR does not exclude serious disease but it makes it unlikely.

2. Once a disease has been diagnosed and is being treated, the change in the ESR gives a rough indication of how the patient is progressing. If the ESR increases the patient is getting worse: if it decreases he is getting better.

RED BLOOD CELLS (ERYTHROCYTES)

Both the red and the white cells are made in the bone marrow by the sequence of events shown in fig. 6.6. During fetal life the liver and spleen both take part in blood cell synthesis and may be able to do so again if during adult life the marrow is damaged. The red cells themselves are flat discs, squashed in a little in the centre of each side (biconcave). They normally live for about 120 days after leaving the marrow.

The marrow in the bones is of two sorts, red and white. The red marrow is where red cell synthesis takes place. The white marrow mainly consists of fatty tissue but like the liver and spleen it can begin to make red cells in an emergency. In adult life most of the red marrow is found in the bones of the skull, vertebral column, thorax and pelvis.

The rate at which the red cells are manufactured depends to a large extent on a hormone produced by the kidneys and known as erythropoietin. The existence of this hormone was first suggested by two clinical observations. In tumours of the kidneys excessive production of red cells frequently occurs while in kidney failure severe anaemia (see next section) is very common. The hormone is at present relatively poorly understood. Its output is increased in anaemia in an effort to return the blood to normal and also during long exposure to low oxygen levels (as in mountain dwellers) when above normal amounts of haemoglobin are required to carry sufficient oxygen to the tissues.

The main function of the red cells is to carry haemoglobin, a red, iron-containing substance which is vital in the transport of oxygen and carbon dioxide around the body and in the maintenance of a steady blood pH. The normal range of haemoglobin concentration is 14–16 g/100 ml blood: women tend to be at the lower end of this range and men at the upper.

The haemoglobin molecule consists of two parts. There is a protein part known as globin and a non-protein, iron-containing part known as haem. As will be discussed later, each complete molecule of heamoglobin (Hb) is capable of combining with four molecules of oxygen.

When red cells become old, they are somehow identified as such and engulfed by the reticulo-endothelial (phagocytic) cells which line blood vessels in the liver, the bone marrow and the spleen. The protein is broken up, the iron is conserved for the manufacture of new red cells and the non-iron part of haem is converted to bilirubin, transported to the liver and excreted in the bile.

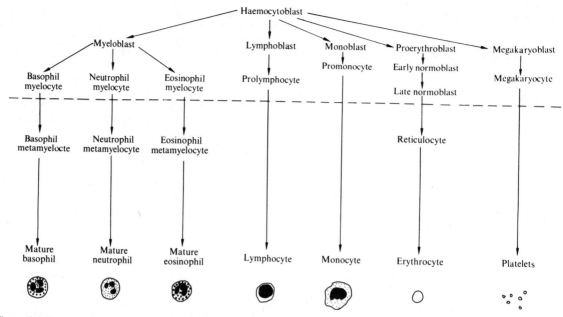

Figure 6.6 The manufacture of red and white blood cells.

ANAEMIA AND NUTRITION

A person is said to be anaemic when the Hb concentration in the blood falls below normal. There are many causes of anaemia including blood parasites such as malaria, abnormalities of haemoglobin, and chronic bleeding because of a peptic ulcer or heavy menstruation. Perhaps the most important cause, however, is a failure to take into the body basic materials required for the manufacture of red cells. Three factors are particularly important.

1. *Iron.* Iron deficiency anaemia is one of the commonest diseases in the world. If too little iron is taken in the food, too little Hb is made to fill the red cells normally. The red cells are therefore small (microcytic) and pale (hypochromic). Normal acid secretion by the stomach is necessary for the absorption of iron from the gut and in the absence of normal gastric function anaemia may occur.

2. *Folic acid.* This is a B group vitamin essential for the manufacture of the cells themselves but not so vital for the making of Hb. Folic acid deficiency therefore produces a shortage of cells and the red cell count is very low. The cells that are made tend to be larger than normal and carry more haemoglobin: they are sometimes known as macrocytes.

3. *Vitamin B$_{12}$ (Cyanocobalamin).* This also is required for cell division and causes a macrocytic anaemia similar to that seen in folic acid deficiency. Anaemia due to vitamin B$_{12}$ deficiency is often known as pernicious anaemia. Almost all diets, except those taken by extreme vegetarians, contain sufficient B$_{12}$ and so a simple dietary deficiency is not the cause. However, B$_{12}$ cannot be absorbed from the gut unless it first combines with a substances known as intrinsic factor which is secreted by the stomach. If intrinsic factor is not secreted, even though there may be plenty of B$_{12}$ in the diet, the vitamin goes straight through to the faeces without being significantly absorbed. Failure of normal intrinsic factor secretion is not uncommon. The disease can be cured by bypassing the gut altogether and giving the vitamin by means of regular injections. The absence of vitamin B$_{12}$ also leads to damage to the spinal cord and peripheral nerves: it may cause behavioural changes. Folic acid deficiency does not cause these defects in functioning of the nervous system.

WHITE CELLS (LEUCOCYTES)

Most of the white cells are concerned with the body's reactions to injury and disease. When a pathologist is asked to look at the white cells in the blood, he usually does two things: he measures the total numbers in a known volume of blood (the normal white cell count is in the region of 4000–8000/mm^3) and also measures the percentages of each type present (a differential white cell count). From measuring the total number and the percentages it is possible easily to work out the numbers of each type. The main types of white cell are:

1. *Neutrophil polymorphs (50–70%)*
These are manufactured in the bone marrow. The name polymorph is short for 'polmorphonuclear' and indicates that the nucleus has a complex shape with a variable number of lobes. The word neutrophil indicates that they

are stained by neutral dyes. About half the mature poly-morphs in the body are not circulating freely in the blood but are stored in the lungs, spleen and marrow. Exercise and adrenaline (epinephrine) bring them out in large numbers as do infections and cell death of any kind, whether due to infection, cancer or injury. These con-ditions therefore all send the polymorph count soaring: a high count is a good indication that inflammation (for example, appendicitis) is taking place somewhere in the body. The neutrophil polymorphs are continually being replaced and their normal life span is probably about five days. They are capable of actively ingesting foreign material and dead or dying body cells by the process known as phagocytosis. Their main functions are there-fore to combat bacterial invasion by engulfing the bacteria and to clear up the debris which results from cell damage. In any part of the body where infection or any other form of inflammation is occurring the blood vessels become sticky. Neutrophils adhere to the sticky capillary walls and then push their way out through the walls into the interstitial fluid where they can operate to combat the disease process. In the course of this many neutrophils are killed and dead neutrophils are a major component of the fluid known as pus.

2. *Eosinophil polymorphs (normally 1–4%)*
These have nuclei which are similar to those of the neutro-phils. However, as their name implies eosinophils are stained red by acid dyes such as eosin and are not stained by neutral dyes. Their functions are not understood. Their numbers do not rise in acute infections and inflam-mations but excessive numbers of eosinophils may be found in some chronic infections such as tuberculosis, in infestations with gut parasites such as worms and in allergic conditions like asthma and hay fever.

3. *Basophils*
These stain with basic dyes. Their function is again un-known but they are rich in histamine and the anti-clotting substance, heparin.

4. *Monocytes (normally 2–8%)*
These are relatively large cells with rounded nuclei. Like the neutrophil polymorphs they can engulf particles by means of phagocytosis. Also like the neutrophils they are attracted to places where infection and inflammation are occurring, but they usually appear on the scene rather later than the neutrophils. They live for several months and they clear up much of the debris left by the neutro-phils. They particularly collect in places where foreign material stays in the body for a long time; surgical suture material and tuberculosis bacteria are examples of par-ticles which tend to cause an accumulation of many monocytes. In such circumstances the monocytes often become transformed into giant cells with many nuclei.

5. *Lymphocytes (normally 20–40%)*
These are smallish cells with rounded nuclei and clear cytoplasm. They play a major part in the immune re-sponse and are discussed in the next section.

THE IMMUNE RESPONSE
Many types of foreign material, when they enter the body evoke what is known as an immune response. Substances which can evoke this response are usually protein in nature and are known as antigens. The response takes five to ten days to develop when a foreign antigen first gains access to the body. It has two main components.

1. *Cellular component.* This is particularly evident when the antigens are carried on the surfaces of foreign cells as in a transplanted organ. It depends on a type of lympho-cyte which is able to migrate to the region where the foreign cells are found and there to destroy them by un-known means.

2. *Antibody component.* Some of the lymphocytes which come into contact with foreign antigens can become transformed into complex cells known as plasma cells which are capable of manufacturing large amounts of a type of protein known as antibody. The antibody has the property of being able to combine with the antigen and so inactivating it. If the antigen is a chemical toxin such as that released by diphtheria bacteria, the combination with antibody can render the toxin non-poisonous. If the antigen is attached to a cell surface the antigen–antibody combination may actually destroy the cell or make it much more susceptible to attack by neutrophils and monocytes.

There are two particularly remarkable features of the immune response. The first is that the response to each antigen is specific. Foreign antigen A will initiate immune response A with manufacture of antibody A: foreign anti-gen B will initiate immune response B with production of antibody B. Antibody B will not combine with antigen A and antibody A will not combine with antigen B. The second feature is that while the first time an antigen enters the body, the defence system takes several days to recognise it and to start the immune response, on the second occasion the response begins within a few hours. The defence system seems to 'remember' how to deal with an antigen it has met before. This explains why many diseases are caught only once: the second time ex-posure to infection occurs, the body is ready and waiting. It also explains the success of vaccination and immunis-ation. The aim of these preventive measures is to expose the body to an antigen similar to one found on a disease-causing organism but which is attached either to a living organism which is harmless or to a dead one. The body thus exposed to the antigen 'learns' about it and is able to mount a very rapid and effective immune response when faced with a real infection.

For cells capable of such complex activities, lympho-cytes are surprisingly simple in appearance consisting of a round nucleus and a scanty amount of clear cytoplasm. They are made in the lymphoid tissue scattered throughout the body in the thymus gland, the gut wall, and the marrow, the spleen and the lymph nodes. The thymus gland, a rather ill-defined mass of tissue situated in the chest near the heart, and the gut lymphoid tissue appear to have particularly important roles in the development and the maintenance of the lymphocyte system. Although much remains to be understood it seems probable that the thymus is primarily responsible for the lymphocytes which take part in the cellular part of the immune response, while the lymphoid tissue in the gut wall is responsible for the lymphocytes which can be transformed into plasma cells and produce antibody. Lymphocytes gain entry to the circulatory system by entering the lymph vessels as they pass through lymph nodes and other types of lymphoid tissue. They are poured into the venous system, mostly via the thoracic duct. They leave the blood again via the capillaries in lymphoid tissue which have unusually permeable walls and thus recirculate continuously between blood and lymph.

BLOOD GROUPS

Suppose samples of blood are taken from a group of people, treated to prevent clotting, and then mixed together in pairs. With some pairs of blood samples nothing will happen but with other pairs the red cells will stick together in large clumps (agglutinate) and the cells themselves may be destroyed. Not until the early years of this century was the research done which explained this phenomenon. It was demonstrated that the human race could be divided into four categories according to the presence or absence of certain proteins on the surfaces of red cells and in the plasma. The red cell proteins are known as antigens or agglutinogens while the proteins in

the plasma are known as antibodies or agglutinins. The four categories are:

Group O. The red cells carry **NO** antigen but there are two antibodies, anti-A and anti-B in the plasma.

Group A. The red cells carry A antigen and there is anti-B antibody in the plasma.

Group B. The red cells carry B antigen and there is anti-A antibody in the plasma.

Group AB. The red cells carry both A and B antigens but the plasma contains neither anti-A nor anti-B antibodies.

The name of a person's group depends on the type of antigen found in his red cells. Whenever red cells carrying A antigen meet with anti-A antibody they will stick together and break up. Similarly whenever red cells carrying B antigen meet anti-B antibody they will stick together. The person's group can therefore be identified if one drop of his blood is mixed with a serum containing anti-B antibody and another drop with anti-A antibody. The mixing is usually done on a microscope slide and clumping together of red cells is looked for. The red cell group can then be readily identified by noting the results and consulting table 6.1.

Table 6.1 Using this table it is possible by mixing red cells with anti-A and with anti-B serum to determine the ABO group to which red cells belong. + means agglutination and can easily be seen because the agglutinated red cells clump together. − means no agglutination.

Cells	Anti-A serum	Anti-B serum
A	+	−
B	−	+
AB	+	+
O	−	−

When blood transfusions are given, it is very important that the two bloods from the patient receiving the blood and the patient giving it should not react together. If the red cells do clump together they may be destroyed (haemolysed) releasing haemoglobin which may block small vessels, particularly in the kidneys and so cause renal failure. In the course of the reaction chemicals known as pyrogens which cause fever and shivering are often released.

When transfusing blood, samples of donor and recipient blood should always be mixed and observed before the transfusion takes place. This is known as crossmatching and is the only certain way of ensuring that there will be no agglutination. In emergencies this may not be possible and the following rules should then be observed:

1. If the recipient's ABO group is known, give blood of the same group.

2. If the recipient's blood group is unknown give group O (universal donor) blood. The O cells carry no antigens and therefore will not be agglutinated by the recipient's

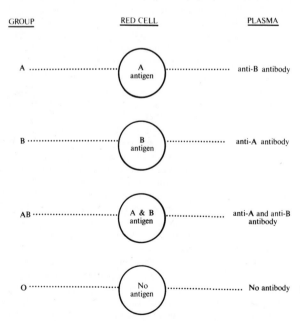

Figure 6.7 Antigens and antibodies present on red cells and in plasma in the four main blood groups.

plasma even if it carries both anti-A and anti-B anti-bodies. The effect of the antibodies in the donor's plasma on the recipient's red cells can usually be ignored because they are so diluted by the recipient's plasma. This may not be true if the antibody concentration in the donor blood is unusually high or in cases where massive transfusion is required.

3. If the recipient's blood group is AB (universal recipient), then blood of any group can be given because there are neither anti-A nor anti-B antibodies in the recipient's plasma.

THE RHESUS PHENOMENON

In 1940 it was found almost by accident that an antibody against rhesus monkey red cells would agglutinate the red cells from most human beings. The human cells were agglutinated because they carried an antigen which became known as the rhesus (Rh) antigen. Those who carried the Rh antigen were known as Rh+. A small proportion of people (around 15% in Europe and North America) did not carry the Rh antigen on their red cells and were known as Rh−. It was soon realised that these findings might explain a not uncommon disease of new born infants. The infants with this disease, when they are born (and often before) suffer from excessive red cell destruction. They become very anaemic and the breakdown of large amounts of haemoglobin leads to jaundice. The bilirubin produced by this haemoglobin breakdown tends to accumulate in the brain and may cause permanent damage.

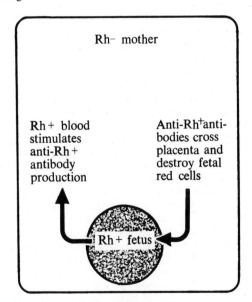

Rh− mother

Rh+ blood stimulates anti-Rh+ antibody production

Anti-Rh+ antibodies cross placenta and destroy fetal red cells

Rh+ fetus

Figure 6.8 The mechanism of Rh disease.

In each case when this happens, the baby and the father are Rh+ while the mother is Rh−. The baby inherits the Rh+ gene from its father. It is also known that the first Rh+ pregnancy in an Rh− woman is almost

never dangerous for the child but that the risk tends to increase progressively with each succeeding pregnancy. In order to understand what is happening it is necessary to appreciate three facts in addition to those already presented.

1. During every pregnancy some cells from the baby's circulation enter the mother's blood. It is unusual for much of the baby's blood to enter the mother until labour begins. Then the uterine contractions inevitably force some of the baby's red cells into the mother's blood.

2. The Rh+ antigen is a foreign protein to an Rh− mother and so she mounts an immune response against it, in the course of which the cells carrying the antigen are destroyed. This immune response takes several days to develop fully.

3. The Rh blood group system differs from the ABO system in that an Rh− mother does not naturally have anti-Rh+ antibodies circulating in her blood.

It should now be possible to understand the following sequence of events which occurs when an Rh− woman has several Rh+ children:

1. In the first pregnancy there are no anti-Rh+ antibodies in the mother's blood. During labour a significant quantity of Rh+ cells from the baby may enter the mother and stimulate the formation of anti-Rh+ antibodies. A few days later these antibodies destroy the baby's red cells present in the mother's circulation but because the baby itself has gone they can have no effect on the child.

2. During a second Rh+ pregnancy the mother will already have been sensitised to the Rh+ antigen. On a second exposure, much smaller quantities of antigen are needed to provoke an immune response and the response occurs much more rapidly. Even if only a very few of the baby's red cells escape into the mother's circulation during the course of the pregnancy, they will stimulate an immediate immune response with the production of large amounts of anti-Rh+ antibody. This antibody may be produced well before the end of pregnancy.

3. If the mother's anti-Rh+ antibodies cross the placenta into the fetal circulation they will destroy the baby's red cells.

4. The baby will become anaemic and the large amounts of bilirubin produced may damage the brain. Brain damage is most likely to occur after birth because during the pregnancy the baby's bilirubin is effectively removed and excreted by the mother. But after birth the baby must rely entirely on its own liver for bilirubin excretion. Even in normal babies, for the first few days the liver may be unable to cope with the bilirubin produced by normal red

cell destruction giving the so-called 'physiological jaundice' which is common a few days after birth. In an Rh+ baby from an Rh− mother the situation is much worse.

The severity of rhesus haemolytic disease depends on the transfer of the baby's red cells to the mother and on the reverse transfer of the mother's anti-Rh+ antibodies to the baby. The degree to which these processes take place varies from pregnancy to pregnancy, accounting for the variation in intensity of Rh disease in different pregnancies in the same mother. Until recently all that could be done was to hope that the baby was not too affected and to give 'exchange transfusions' after birth to affected babies. In an exchange transfusion some of the baby's own blood is removed and an equal quantity of fresh donor blood is put back. This has the double effect of lowering the bilirubin levels and combating the anaemia. Recently two new techniques have considerably improved the outlook in Rh disease.

1. It is now possible to transfuse a baby even while it is still in the mother's uterus. Under X-ray control a needle is passed into the baby's peritoneal cavity and red cells are injected in. The red cells are, perhaps surprisingly, rapidly absorbed into the blood stream.

2. The initial stage in Rh disease depends on the entry of the baby's red cells into the mother in significant amounts, usually at the time of the birth of the first Rh+ infant. If these red cells could be promptly destroyed, they would not be able to stimulate the mother's immune response and so sensitisation could not occur. It has now been proved that the baby's red cells which appear in the mother can very rapidly be destroyed by injecting into the mother at the time of delivery anti-Rh+ antibodies. These rapidly haemolyse any fetal Rh+ cells and prevent the development of the immune response. By injecting anti-Rh+ antibody immediately after delivery in every pregnancy in an Rh− woman it should be possible to prevent maternal sensitisation almost entirely and so to eliminate rhesus disease in any population where reasonable standards of medical care are available to all.

BLOOD CLOTTING

As everyone knows, when blood is taken from a blood vessel and put into a container it does not remain liquid but clots. A normal clotting mechanism is obviously essential if the body is to respond to injury normally.

The sequence of events which occurs in clotting is very complex in detail but relatively simple in outline. Only the outline will be given here. The process may be started off in two ways:

1. If the blood comes into contact with an abnormal surface clotting may begin. An abnormal surface may arise in the body because of oxygen lack which may occur when the blood in a vessel ceases to flow and becomes stagnant.

2. If products of damage to body tissues gain access to the blood the process of clotting is initiated. This may also occur during injury.

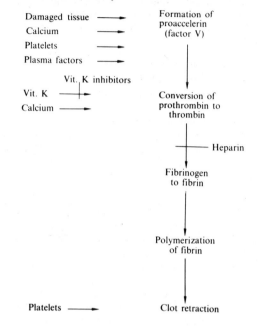

Figure 6.9 An outline of blood coagulation.

Both preliminary phases end up with the formation of a substance called thromboplastin (sometimes also known as active factor V). Calcium, and the non-nucleated formed elements of the blood known as platelets are both essential for these preliminary stages leading to thromboplastin formation.

Thromboplastin is an enzyme which can act an a protein found in the blood known as prothrombin. Prothrombin is manufactured by the liver and vitamin K is essential for its synthesis. In the presence of calcium, thromboplastin converts prothrombin into another active enzyme known as thrombin. Thrombin then acts on another plasma protein, fibrinogen, to form fibrin. Fibrin consists of long threads which mesh together and trap red cells, white cells, platelets and plasma. The trapped platelets then release substances including serotonin or 5-hydroxytryptamine which cause the fibrin threads to contract, making the clot firmer and expelling the clear yellow fluid known as serum: this process is known as clot retraction. The difference between plasma and serum is often not understood. Plasma is the fluid part of the blood, excluding the formed elements: it can be obtained by centrifuging unclotted blood. Serum is identical to plasma apart from the fact that the clotting factors present in the plasma have been removed.

Nearly twenty factors have now been identified as being necessary for the normal clotting of blood and more seem to be discovered every year. Many diseases are known in which one of these factors is congenitally absent. The most famous is haemophilia due to absence of factor VIII or anti-haemophilic globulin. Haemophilia became famous because it is relatively common, because it occurred in some of the European royal families and because, since the normal gene is carried on the X chromosome and since females have two Xs but males only one, it is very much commoner in males than females. For a female to have haemophilia both her X chromosomes must be abnormal: females with one abnormal X chromosome do not show the disease but they can pass on the abnormality and the disease to their sons. The main features of haemophilia and of most of the other clotting factor defects are a failure of wounds to stop bleeding and a tendency for bleeding to occur into joints. During normal exercise joint blood vessels often become damaged but the blood clots at once and cause no trouble. In haemophiliacs the bleeding from such minor injuries may continue for long periods.

When blood is withdrawn from one person in order to give a transfusion to another, it is obviously essential to stop the blood clotting in the bottle. This is usually done by withdrawing ionic calcium from the donor blood by means of a chemical reaction with citrate, oxalate or EDTA ('sequestrene'). Since free calcium ions are required for several stages in the coagulation process, these substances can effectively prevent clotting

Anticoagulants
The ease with which blood clots tends to be considerably increased after tissue damage such as occurs during a surgical operation or a heart attack. When patients are in bed, the blood in the leg veins tends to stagnate. The combination of these two factors may lead to abnormal and dangerous clotting within the blood vessels themselves. Such patients may therefore need treatment to reduce the clotting activity of the blood. They can be given two different types of drug.

1. *Heparin.* This was originally isolated from leeches which suck blood and which need to prevent it clotting so that they can feed properly. It is also found in the basophil leucocytes. It is digested by the human gut and so must be given by injection. It acts instantaneously, interfering with clotting by blocking the conversion of prothrombin to thrombin. Its action can be instantly reversed by injecting the protein protamine which is found in fish sperm: the protamine acts by combining with the heparin.

2. *Oral anticoagulant drugs.* There are many varieties of these, all acting by inhibiting the action of vitamin K in the liver. Some of the substances occur naturally in plants and can cause bleeding when eaten by farm animals. They prevent the manufacture of prothrombin and other clotting factors. They do not interfere with the actions of clotting factors already present in the blood but as these factors are naturally used up and destroyed they are not replaced. The oral anticoagulant drugs therefore take several days to act. When their intake is stopped the liver takes some time to restore the levels of clotting factors to normal.

HAEMOSTASIS
This is the stopping of bleeding from damaged blood vessels. It is often thought that clotting (coagulation) is the only process of importance in haemostasis but this is not true. Even if clotting is defective, bleeding may stop entirely normally from small wounds of the skin: even if clotting is normal, other deficiencies may lead to prolonged bleeding from small wounds. The cessation of bleeding depends on three distinct but interlinked processes.

1. When small blood vessels are damaged, the injury stimulates the contraction of the muscular tissue in their walls. This muscle spasm tends to close off the vessel.

2. The damage stimulates platelets to stick together to form a plug over the hole which prevents the leakage of blood. This process is known as conglutination.

3. Finally the blood in the damaged vessel clots to form a firm plug which will control the escape of blood until healing has taken place.

With small wounds the first two processes alone may be able to stop bleeding but with large wounds haemostasis cannot occur unless clotting is normal.

THE SPLEEN
Although most people can survive entirely satisfactorily without a spleen, the organ nevertheless does perform important functions. It is rich in lymphoid tissue. It contains blood sinuses which are lined by phagocytic reticulo-endothelial cells which can engulf particles from the blood.

The main functions of the spleen are:

1. Red cell manufacture occurs in the spleen in the fetus. In children or adults whose bone marrow has been damaged or destroyed, the spleen may start making red cells again.

2. Removal of unwanted particles from the blood. The spleen is important in the destruction of ageing red and white cells and platelets.

3. The lymphoid tissue like lymphoid tissue everywhere plays an important part in the immune response.

4. Red cell storage. This is known to be of importance in carnivorous animals. To meet an emergency the spleen contracts vigorously pouring out a large store of red cells held within the sinuses into the general circulation. The importance of this is not known in man.

7 Circulation

The circulation forms the body's transport system which by pumping blood puts every part of the body in communication with every other part. First it is important to have an outline of the structure of the system.

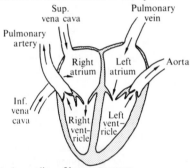

Figure 7.1 An outline of heart structure.

The human heart contains four chambers. Blood from the peripheral areas of the body apart from the lungs is returned by the veins to the thin-walled right atrium. From there it passes to the right ventricle which pumps it out along the pulmonary artery to the lungs where it is oxygenated. The pulmonary veins return the oxygenated blood to the left atrium from which it passes into the left ventricle which pumps it out into the aorta and so to the peripheral parts of the body. The period when the heart is contracting and pumping blood out is known as systole. The period when it is relaxing and filling with blood is known as diastole. The heart is an effective pump because the entries to and exits from the ventricles are controlled by non-return valves. When the ventricles contract the valves guarding the channel between the atria and ventricles (tricuspid valve on the right and mitral valve on the left) close to prevent the backward flow of blood. The valves controlling the exits from the ventricles to the pulmonary artery on the right and aorta on the left open to let the blood out. The valves in the pulmonary artery are known as pulmonary valves and the ones in the aorta as the aortic valves.

As it passes round the body, the blood first goes through arteries, then arterioles, then capillaries, then

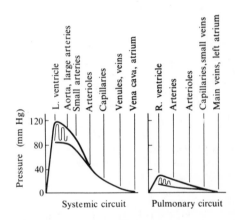

Figure 7.2 The pressures in the various blood vessels in a normal individual.

venules and then veins before getting back to the heart again. All the blood vessels except the capillaries have four layers in their walls:

1. An innermost lining of endothelial cells, one cell thick (tunica intima). This layer is the only one present in the capillaries. It allows all the constituents of the blood except the cells and plasma proteins to pass through it freely and it is therefore in the capillaries that the exchange between blood and tissues takes place. The other three layers are relatively impermeable to the passage of plasma and its constituents.

2. An inner layer of connective tissue fibres, mainly arranged longitudinally.

3. A middle layer of mixed connective and muscle fibres (tunica media), arranged circularly.

Both elastin fibres (which stretch more easily than rubber) and collagen fibres (which are resistant to stretch) are found in the connective tissue.

In arteries all three outer layers are well-developed but the middle layer contain much more elastic tissue than muscular tissue. As the arteries divide and decrease in diameter, the inner and outer layers of connective tissue become much thinner and the proportion of muscle in the middle layer increases giving the vessels known as arterioles. The arterioles give way to the capillaries, an intricate network of minute vessels with walls only one cell thick. The capillaries coalesce to give venules which in turn combine to give veins which have much thinner walls than the corresponding arteries. Table 7.1 gives information about the number, diameter, cross-sectional area and volume of the blood vessels in the gut of a mammal. It is important to note that although the capillaries

Table 7.1 The measurements of blood vessels to the gut in the dog showing the changes in size and number of vessels at different levels (data of Mall).

Vessel	Diameter (mm)	Number	Total cross-sectional area (cm²)
Aorta	10	1	0·8
Main arterial branches	1	600	5·0
Arterioles	0·02	40 million	125
Capillaries	0·008	1,200 million	600
Main venous branches	2·4	600	27
Vena cava	12·5	1	1·2

are so small, there are so many of them that their total cross-sectional area is this single vascular bed is about six hundred times that of the aorta.

The Muscle Pump
The heart is not the only pump in the body. The veins also contain non-return valves. If a vein is compressed by

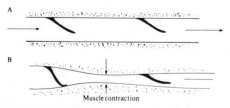

Figure 7.3 The operation of the valves in the veins. A. At rest. B. During muscular contraction.

muscular activity the blood is squeezed out and has no alternative but to move towards the heart. At rest the muscle pump is ineffective and all the work of pumping the blood around the body must be done by the heart. But during exercise when the muscles are repeatedly contracting and relaxing, quite a high proportion of the work of the circulation may be carried out by this muscle pump.

Thoracic Pump
Another factor which can help in the return of venous blood to the right side of the heart is the so-called thoracic pump. During inspiration as the diaphragm is pulled downwards and the chest wall expands, pressure in the abdomen rises while pressure in the thorax falls. These pressure changes pump blood up from the abdominal great veins into the thorax.

THE HEART
HEART MUSCLE
When seen under a microscope heart muscle appears to be striated but it differs in several ways from skeletal muscle. It consists of branching fibres which are

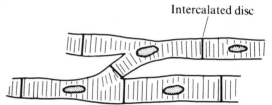

Figure 7.4 Heart muscle fibres.

extremely closely attached to one another by structures known as intercalated discs. The effect is to link all the fibres very closely together both structurally and functionally. In behaviour heart muscle differs in two major ways from skeletal muscle.

1. It has a very long refractory period which lasts longer than the period of contraction. In skeletal muscle the refractory period lasts not much more than 0·001 sec (1 millisecond) while the contraction lasts from 0·2–0·5 sec. This means that if two nerve impulses follow one another

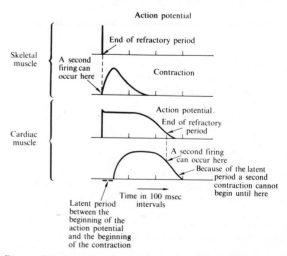

Figure 7.5 A comparison of the action potentials and subsequent contractions in skeletal and cardiac muscle. In skeletal muscle there is virtually no latent period between the action potential and the beginning of the contraction but in cardiac muscle the latent period is relatively long.

very quickly the muscle has no time to relax and the contractions add on to one another. It would be quite disastrous if this sort of thing happened with the heart. The heart depends for its action on a strong contraction which pumps out the blood and which is always followed by a period of relaxation when the heart can fill with blood again ready for the next contraction. It is essential that contractions should not be able to follow one another too quickly: this is achieved by the very long refractory period, longer than the contraction itself, which ensures that the muscle cannot be excited again until it has fully relaxed.

2. Even in the total absence of nerves, heart muscle strips contract spontaneously and rhythmically. Skeletal muscle normally contracts only in response to a nervous impulse: in the absence of such an impulse it remains relaxed. But even if all the nerves to the heart are cut, or even if strips of heart muscle are bathed in an appropriate oxygenated solution, the heart muscle fibres spontaneously in a rhythmic manner. The use of isolated strips taken from different parts of the heart reveals that heart muscle may be divided into three broad categories according to its origin and behaviour.

    a. Muscle from the so-called 'pacemaker region' (the area in the right atrium known as the sino-atrial or SA node) beats fastest, with a spontaneous rate of about 70–80 beats/min.

    b. Muscle from other parts of the atria beats spontaneously at about 50–70 beats/min.

    c. Muscle from the ventricles beats slowest at a natural rate of 20–40 beats/min.

CONDUCTION OF THE CARDIAC IMPULSE

As in skeletal muscle, contraction of cardiac muscle is fired off by an impulse (action potential) spreading over the surfaces of the muscle fibres. In skeletal muscle, the muscle impulse must be fired off by a nerve impulse reaching the neuromuscular junction. In cardiac muscle the muscle impulse arises spontaneously without nervous action.

    There is a group of muscle fibres in the right atrium known as the sino-atrial (SA) node. The SA node is sometimes called the pacemaker since its fibres naturally beat faster than other heart muscle fibres and because it 'calls the tune' forcing the other parts of the heart to follow its rhythm under normal circumstances.

    When an impulse arises spontaneously in the SA node it spreads out along the atrial muscle fibres so making the atria contract. The atria and the ventricles are almost completely separated from one another by fibrous connective tissue which does not conduct impulses. The only conducting connection between the atria and ventricles is a small piece of specialised muscle tissue known as the atrio-ventricular (AV) node. Here the conducting fibres become very thin and in consequence the impulse travels very slowly from the atria to the ventricles. This ensures that the contraction of the atria is virtually completed by the time the impulse reaches the ventricles.

On the ventricular side there is a specialised system of muscle fibres known as Purkinje fibres which conduct the impulse extremely rapidly to all parts of the ventricle.

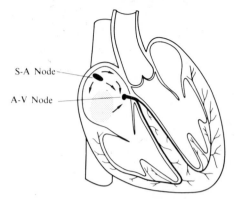

Figure 7.6 The conducting system of the heart.

They are first gathered together in the bundle of His but this soon divides into right and left bundle branches which go to their respective ventricles. The Purkinje fibres conduct the impulse rapidly so that it arrives almost simultaneously at all parts of the ventricles. This enables all the ventricular fibres to contract vigorously together.

THE HEART RATE

The natural spontaneous rate of beating of the pacemaker may be altered in four main ways:

1. Sympathetic nerves to the SA node release noradrenaline which increases the rate of beating.

2. Vagal (parasympathetic) fibres to the SA node release acetyl choline which slows down the rate of beating.

3. Circulating adrenaline which has been released from the adrenal medulla acts on the SA node to increase the heart rate.

4. Distension of the right atrium due to overfilling with blood causes the SA node fibres to be stretched and this makes them contract more vigorously. This may be one of the factors which helps to account for the fast heart rate in the condition known as congestive cardiac failure when the heart is greatly dilated.

The Origin of the Spontaneous Heart Beat

The reason for the spontaneous beating of cardiac muscle is now at least partly understood. Understanding has come as a result of intracellular recordings of the electrical activity of cardiac muscle fibres. You may remember that with skeletal muscle at rest, the potential difference between the inside and outside of the fibre (the membrane potential) remains constant at about -70 mV inside. Only when a nerve impulse releases acetyl choline does an action potential develop and trigger a contraction. After the action potential is over the membrane potential returns to −70 mV and stays there until another nerve impulse comes along. The action potential and the corresponding refractory period in skeletal muscle are both of over within about 1·5 msec.

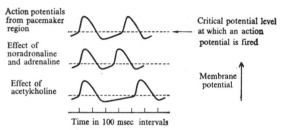

Figure 7.7 Action potentials recorded intracellularly from the pacemaker region of the right atrium illustrating the actions of acetyl choline, adrenaline and noradrenaline.

Heart muscle is quite different. Recordings have shown first of all that instead of rising quickly and returning quickly to the resting level, the membrane potential rapidly becomes positive and stays there for 100–400 msec, depending on the species and on the part of the heart from which the muscle is taken. In general, action potentials in the atria are shorter than those in the ventricle but in both cases the action potential, and hence the refractory period, lasts longer than the contraction which is fired off when the action potential begins. This means that the heart muscle is refractory until relaxation has taken place: two successive contractions cannot normally add together and the heart cannot go into a tetanic state. This is extremely important as the action of the heart depends just as much on filling during relaxation as an emptying during contraction. The second major point revealed by intracellular recording is that in the absence of nerve activity the membrane potential of heart muscle is *not* stable. After an action potential has been completed and the membrane has been repolarised (to a level usually in the region of −50 to −70 mV depending on the particular type of muscle) it does not remain stable. Instead immediately it begins to be slowly depolarised again. When this level of depolarisation reaches a certain critical point, known as the threshold, another action potential is initiated. Thus automatically a train of action potentials is fired off by the membrane and these action potentials produce a rhythmic series of contractions. Intervals between action potentials depend on the rate at which the spontaneous depolarisation takes place.

It has also been shown that acetyl choline, noradrenaline and adrenaline alter the rate of beating of the heart by altering the rate at which the depolarisation occurs. Adrenaline and noradrenaline seem to make the membrane less stable than usual and the rate of depolarisation is more rapid. This means that the threshold is reached more quickly, the interval between action potentials become shorter and the heart rate increases. Acetyl

choline, in contrast, stabilises the membrane. Depolarisation therefore takes place more slowly, the interval between action potentials increases and the heart rate slows.

THE CARDIAC CYCLE
The description of the cardiac cycle which follows applies to the left side of the heart but the right side behaves in a similar manner.

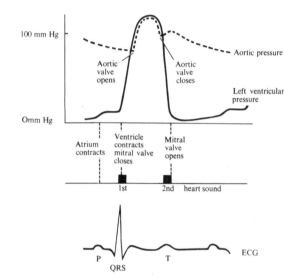

Figure 7.8 The main events of the cardiac cycle.

1. As the ventricle relaxes after its last beat, the pressure of blood within it falls rapidly. When the pressure in the ventricle becomes lower than that in the atrium the mitral valve opens and blood rushes into the empty ventricle. As the ventricle becomes full, the rate of filling slows down. It receives a boost when the atrium contracts so forcing a little extra blood into the ventricle. It should be noted, however, that the ventricles can fill reasonably well even in the absence of atrial contraction.

2. Because of the delay at the AV node, the ventricle contracts well after the atrium. As soon as the ventricular contraction begins, the pressure of the blood in the ventricle rises sharply. As soon as the pressure in the ventricle becomes higher than the pressure in the atrium the mitral valve closes. As pressure rises higher, the pressure in the ventricle becomes higher than the pressure in the aorta and this forces open the aortic valves. Blood rushes out into the aorta. When a person is resting, the ventricle pumps out only about half the blood it contains, the remainder staying in the ventricle at the end of the beat. During exercise the ventricle beats more vigorously and less blood is left behind at the end of each beat.

3. As the ventricle relaxes, pressure inside it falls below the pressure in the aorta. When this happens the aortic valve snaps shut, so preventing any backflow from the

aorta into the heart. As relaxation continues, the pressure in the ventricle falls below that in the atrium. The mitral valve opens to let the blood which has been accumulating in the atrium flow into the ventricle.

Sounds from the Heart and Arteries

Liquids flowing along tubes may travel in two ways, by means of laminar flow or by means of turbulent flow. The difference between the two is shown in fig. 7.9 which

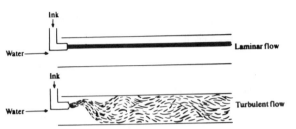

Figure 7.9 Turbulent and laminar flow.

shows a fine jet of ink being injected into a tube. When flow is laminar, the ink does not become mixed with the water but remains as a separate layer or lamina, hence the name laminar flow. The whole force of the flowing liquid is directed to pushing it onwards and none is wasted on side-to-side movement which would mix up the layers and force the liquid to bang against the sides of the tube.

Turbulent flow is quite different and its nature can be guessed at from its name. Immediately on entering the tube the ink is thoroughly mixed up with the water. Much of the force of the flowing liquid is wasted by side-to-side movement of the water and in impacts upon the walls of the tube. Thus the liquid flowing along the tube rapidly becomes thoroughly mixed.

Three main factors determine whether flow will be laminar or turbulent.

1. The diameter of the vessel. Turbulent flow becomes much more likely as a tube becomes narrower.

2. The speed of flow. The likelihood of turbulence increases as the speed of flow increases.

3. The presence of irregularities in the vessel wall or of sudden changes in the direction of the tube.

If a stethoscope is placed over a heart or blood vessel in which turbulent flow is occurring, a rushing sound or 'murmur' can be heard. In contrast, laminar flow is completely silent. This is because the side-to-side movement of the fluid in turbulent flow sets the walls of the vessel vibrating. These vibrations are picked up by the stethoscope as sounds. A murmur therefore can occur whenever flow is turbulent.

The classical heart sounds (which are *not* murmurs) originate as vibrations in various parts of the heart,

notably the valves. Two heart sounds can be clearly heard in any normal person by anyone who picks up a stethoscope for the first time. They are often described as sounding like lub-dup. The first sound is due to the closing of the mitral and tricuspid valves: the closing of the two is normally synchronous and only one sound can be heard. The second sound is due to the closure of the aortic and pulmonary valves. These events may be synchronous or, even in healthy people, the second sound may be slightly split because the contraction of the left ventricle is completed a fraction of a second before that of the right. It is thus apparent that the period between lub and dup corresponds to the period of ventricular contraction (systole) while the period between dup and lub corresponds to ventricular relaxation (diastole).

In addition to the heart sounds, various murmurs due to turbulent flow of blood may arise in the heart:

1. Those occurring between the first sound (lub) and the second sound (dup) arise during ventricular contraction and must therefore be associated with the turbulent expulsion of blood from the ventricle.
   a. If the tricuspid or mitral valves fail to close properly (valvular incompetence) there may be an abnormal backward flow of blood from ventricle to atrium.
   b. Turbulence may simply be caused by a very rapid flow of blood through normal valves. This may occur in the smaller vessels of children, especially during exercise, and has no pathological significance.
   c. The pulmonary or aortic valves may fail to open properly (valvular stenosis). Turbulence occurs as the blood is forced out through the abnormally narrow opening.

2. Murmurs occurring between the second and first sounds, during ventricular relaxation, must be associated with the flow of blood into the ventricles.
   a. If the aortic or pulmonary valves are incompetent they may not close properly and backward flow from the pulmonary artery or aorta into the ventricles may occur.
   b. Stenosis of the tricuspid or mitral valve may cause a murmur as blood flows through the narrowed channel.

THE ELECTROCARDIOGRAM (ECG or EKG)

When the heart muscle contracts, it sets up small electrical currents which are conducted through the body fluids. These currents may be detected by electrodes connected to the skin at appropriate points and connected to a sensitive recording instrument known as an electrocardiograph. The electrodes may be attached to the arms and legs and to the chest. By recording from different parts of the body the contribution made by the activity of different parts of the heart may be emphasised.

A normal ECG from 'lead II' (electrodes on the right arm and left leg) is shown in fig. 7.10. Five waves are conventionally described. The P wave is associated with the contraction of the atrial muscle fibres. The Q, R and S waves occur very close together and are often known as the QRS complex. They are associated with the start of ventricular muscle contraction. The T wave is associated with the end of ventricular contraction.

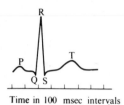

Figure 7.10 A normal ECG from lead II.

As a result of long experience, changes in the ECG pattern have come to be associated with particular cardio-vascular diseases such as coronary thrombosis and hypertension. The ECG is therefore now a very important technique in diagnosis.

ARRHYTHMIAS

When the heart is beating entirely normally it is sometimes said to be in sinus rhythm because it is following the rhythm set by the sino-atrial node. Deviations from this normal rhythm are called arrhythmias. There are many varieties but some of the more important ones are as follows:

1. *Ectopic beats*. These are additional beats occurring outside the normal rhythm. They may arise anywhere in the heart, either in the atria or the ventricles because of some unusual irritability of part of the muscle. They may be of no pathological significance or they may indicate serious disease.

2. *Atrial flutter*. The atrial contractions are regular, but very rapid (often over 200/min). Since the refractory period of ventricular muscle is longer than that of atrial muscle, the ventricles cannot contract more than about 200 times per minute. This means that some of the atrial impulses arrive at the AV node while the ventricle is refractory. The ventricle usually beats relatively regularly but one in every two, three or four of the atrial impulses fails to get through.

3. *Atrial fibrillation*. The rate of beating of the atrial fibres is extremely rapid (often over 400/min) and is no longer regular. The AV node is therefore bombarded with hundreds of irregular impulses every minute. Only a few get through and the ventricular beat is very irregular. Relatively common causes of atrial fibrillation are stenosis of the mitral valve and overactivity of the thyroid gland although the precise mechanisms are not well understood. Perhaps surprisingly, neither atrial flutter nor atrial fibrillation reduce the cardiac output to levels which are immediately dangerous.

4. *Ventricular fibrillation*. The ventricular muscle fibres beat very rapidly and in a totally uncoordinated way so that the ventricle as a whole cannot contract. The patient dies quickly because cardiac pumping falls to zero unless defibrillation is brought about rapidly. This may sometimes be achieved by giving the patient a sharp blow on the chest. More usually it is necessary to give a strong electric shock to the heart through special electrodes. This causes all the heart muscle fibres to contract simultaneously and subsequently to become refractory simultaneously. On recovery from the refractory state the heart may begin to beat normally. The commonest cause of ventricular fibrillation is probably coronary thrombosis, but exessive amounts of adrenaline can sometimes initiate it, especially in patients who are anaesthetised.

5. *Heart block*. There are many types of damage to the conducting system. The simplest to understand is complete heart block in which the AV node ceases to function. The atria continue to beat at their normal rate of about 70/min but no impulses can get through to the ventricles which therefore beat at their normal isolated rate of around 30/min.

IONS AND THE HEART

If the heart is to beat normally, the blood passing through it must contain precisely the right concentrations of certain ions and in particular of potassium and calcium.

Raising the potassium concentration weakens contractions and may cause arrhythmias or even stop the heart. Lowering the potassium concentration makes the heart abnormally excitable and may also cause arrhythmias. A high potassium concentration tends to occur in renal failure when the kidneys cannot excrete potassium ions normally. Low potassium concentrations are most common in patients who are being treated with diuretics to increase urine flow, for many diuretics increase urinary potassium loss. Low potassium levels are particularly dangerous in the presence of digitalis and when the two are present together arrhythmias occur frequently.

High concentrations of calcium can stop the heart in systole. This could occur during overactivity of the parathyroid glands when there is excessive loss of calcium from the bones into the blood but in practice is uncommon.

THE PULSE

When the left ventricle pumps, the blood initially enters the aorta far more rapidly than it can leave via the major arteries. The aorta therefore stretches in order to accommodate the extra blood. During diastole, as the blood flows away, the aorta returns to its normal size. This stretching sets up a pulse wave in the arterial walls: this wave has nothing to do with the actual flow of blood along the arteries and travels much more rapidly. The speed of travel of the pulse wave increases as the arterial walls become stiffer with advancing age. In childhood it travels at about 4 m/sec and in old age at about 10 m/sec.

The pressure difference between the peak pressure reached in the aorta during systole and the lowest pressure during diastole is known as the pulse pressure. The greater this pressure the more powerful will be the wave set up and the stronger the pulse will feel. The following are some important factors which alter the strength of the pulse:

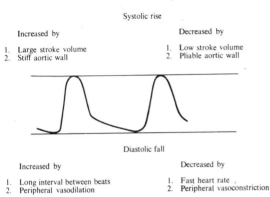

Systolic rise

Increased by
1. Large stroke volume
2. Stiff aortic wall

Decreased by
1. Low stroke volume
2. Pliable aortic wall

Diastolic fall

Increased by
1. Long interval between beats
2. Peripheral vasodilation

Decreased by
1. Fast heart rate
2. Peripheral vasoconstriction

Figure 7.11 Factors which affect the pulse pressure.

1. The amount of blood ejected by the ventricle in one beat (the stroke volume). The more blood is pumped out, the greater will be the pressure rise and the stronger will be the pulse.

2. The stiffness of the aorta. The stiffer the aorta, the greater will be the pressure rise and the stronger will be the pulse.

3. The rate of outflow from the aorta. Blood may leave the aorta rapidly if the aortic valve is incompetent so allowing blood to pass back into the ventricle or if the peripheral arterioles are open wide so that blood can flow into the capillaries very rapidly. This rapid outflow of blood will lower the diastolic pressure and increase the pulse pressure. If there is marked vasoconstriction such constriction of the arterioles such as may occur after a haemorrhage, the rate of outflow from the aorta will be reduced and the pulse pressure will be low.

4. The rate of beating of the heart. The faster the heart is beating, the less chance is there for pressure to fall between beats and the lower will be the pulse pressure.

THE BLOOD PRESSURE

Undoubtedly the most accurate way to measure arterial pressure is to stick a needle connected to a sensing device into a major artery. For most purposes this is impractical and an indirect instrument, the sphygmomanometer must be used.

A wide rubber bag is wrapped around the upper arm at about the level of the heart. The rubber bag is connected to a manometer for measuring the pressure within it and to a simple pump by which it can be inflated. Before the bag is blown up, the brachial artery is detected at the elbow by palpation and the stethoscope bell is placed over it. Nothing can be heard because the flow in the artery is smooth and laminar and not turbulent. The bag is then blown up to a pressure of around 250 mmHg which is well above the systolic pressure in most individuals. This high pressure will stop blood flowing through the brachial artery since it is well above the systolic pressure in most individuals: no sounds can be heard through the stethoscope. While listening through the stethoscope, the pressure in the bag is then slowly released by means of a valve. Most people can hear three types of sound:

1. When the bag pressure falls just below the systolic pressure, a spurt of blood will force open the artery for a moment as the pressure in the artery reaches its peak. The spurt of blood causes rapid, turbulent flow and a sharp tapping sound occurs each time the ventricle ejects blood into the aorta.

2. As long as the bag pressure is above the diastolic pressure, the artery will be closed during each diastolic period when the pressure within it falls. When the bag pressure falls just below diastolic pressure, blood will flow through the artery during diastole as well as systole. The sounds then become less clear and distinct and tend to run into one another. This change is known as muffling.

3. The bag may partially block the artery during diastole until much lower pressures are reached. This may cause turbulence of flow and a sound which does not disappear until the bag pressure falls much lower.

It is widely agreed that the appearance of the sharp tapping sound is a good indication of the true systolic pressure. There is not nearly so much agreement about what indicates the diastolic pressure: some authorities think that the muffling indicates diastolic pressure while others use the disappearance of sounds. Fortunately the muffling and disappearance occur quite close together in most people and for practical purposes either one can be used, provided that it is used consistently.

THE CARDIAC OUTPUT

The cardiac output may be defined as the volume of blood pumped out by the left ventricle in one minute. It depends on the rate of the heart (the number of beats/min) and on the stroke volume (the amount of blood pumped out in one beat).

In a normal adult the cardiac output is usually in the region of 4–5 L/min. In exercise it may increase enormously to 30 L/min or more. In a resting person the actual distribution of the blood to the various organs of the body is shown in table 7.2. During exercise the blood

Table 7.2 Approximate distribution of the cardiac output at rest. During exercise kidney and gut flow are greatly decreased and muscle flow is greatly increased.

Kidney	20–25%
Liver and gut	30–35%
Muscle	10–20%
Head	15–20%
Heart	3– 7%
Other organs	10–20%

flow to the gut and to the kidneys is cut down considerably and more blood is diverted into the working muscles.

The cardiac output may be altered by changing either the heart rate or the stroke volume, or both simultaneously. The ways in which the heart rate may be changed have already been mentioned earlier in this chapter. The next section is concerned with stroke volume.

Stroke Volume

It is important to realise that in a normal resting individual the ventricles do not normally push out with each beat all the blood that they contain. If a ventricle contains 80 ml blood at the end of the filling phase and before contraction begins, only 40–50 ml may be pumped out with each beat leaving 30–40 ml of blood behind. There are two main ways in which the stroke volume may be altered:

1. The maximum heart size during the cardiac cycle (i.e. the heart size when the ventricles are full just before contraction begins) may remain unchanged but the ventricles may beat more vigorously during each systole. This means that a greater proportion of blood is pumped out. In the terms of the example just given, the left ventricle may still contain only 80 ml when filling has been completed, but instead of pumping out 40–50 ml, it will pump out 60–80 ml. The ventricles may be made to beat more vigorously in two main ways:

a. By the action of noradrenaline released from the sympathetic nerves to the heart.

b. By the action of adrenaline released from the adrenal medulla into the circulating blood.

2. When stretched up to a certain limit, all types of muscle contract more vigorously than when they are not stretched. If the stretching goes beyond the limit, however, the ability of the muscle to contract fails rapidly. Heart muscle is no exception to this general rule as was first demonstrated by Frank in Germany and Starling in England. If the ventricular muscle becomes stretched because the ventricle contains a larger amount of blood at the end of the filling phase, then the ventricle will contract more vigorously and expel a greater than usual percentage of the blood it contains. This phenomenon is commonly given the name of 'Starling's Law of the Heart' This mechanism is important in two main situations:

a. At the beginning of exercise when many of the skeletal muscles start contracting blood is squeezed out of the veins because of the muscle pump. This means that there is a sudden increase in the venous return of blood to the right heart. This may temporarily stretch the right ventricle and make it contract more vigorously.

b. When a heart is failing for any reason, it can no longer pump out the normal amount of blood so easily. This means that more blood tends to be left in the ventricles at the end of systole. During diastole a normal amount of blood comes back from the heart to fill the ventricles and the result is that at the end of the filling phase the ventricles contain more blood than normal. The heart as a whole becomes fuller and larger than it was before. The stretch of the ventricular muscle temporarily restores the ability of the heart to pump out a normal stroke

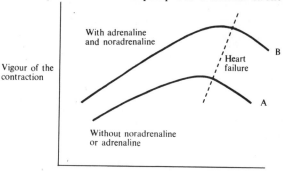

Figure 7.12 The relationship between the volume of the ventricle and the amount of work it can do.

volume. As the heart progressively fails it therefore becomes bigger and bigger (because it contains more blood at the end of the filling phase) in an effort to pump out a bigger stroke volume. The stretching continues until the limit is reached where further stretching causes a decrease in contractile vigour rather than an increase. When this point is reached, unless effective treatment is carried out there is a rapid decline in cardiac output and the person tends to die.

In practice, of course, the two ways of altering stroke volume, by stretching the ventricular muscle and by stimulating it with adrenaline and noradrenaline, interact in a complex way. Perhaps the simplest way of looking at this is to consider the effects of stretching the ventricular muscle under different dates of constant stimulation by the sympathetic system. Fig. 7.13 shows what might

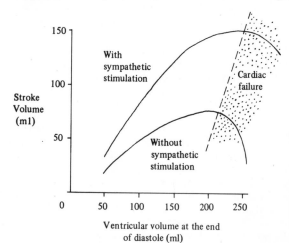

Figure 7.13 The effects of ventricular volume and sympathetic activity on stroke volume.

happen under three arbitrary states of sympathetic stimulation, normal, low and high. When sympathetic stimulation is high, the stroke volume is higher than when sympathetic stimulation is normal at every degree of ventricular stretch although the overall pattern of the curve is similar. When sympathetic stimulation is low then stroke volume is below normal at every point. Stroke volume may be altered either by moving along a curve, or by moving from one curve to another or by both methods simultaneously.

THE PERIPHERAL CIRCULATION

Once the blood leaves the left ventricle it is carried to the tissues by arteries, arterioles and capillaries. On leaving the capillaries it is taken back to the right heart by the venules and veins. All these vessels constitute what is known as the peripheral circulation.

ARTERIES

These are tough walled tubes whose main function is simply to act as conduits through which the blood is carried to the tissues. Their walls are rich in elastic and muscular tissue and this makes them very strong. If an artery is cut, the muscle in its wall contracts very vigorously and this contraction may even succeed in closing quite a large vessel.

ARTERIOLES

The function of the arteries is simply to act as channels carrying blood to the organs. They do not regulate the amount of blood going to a particular organ and they are not under minute by minute control by nerves or hormones. The arterioles are quite different. They are formed as a result of the repeated branching of arteries and they are small vessels very rich in smooth muscle. This smooth muscle can be controlled by nerves, by hormones and by some other chemicals. It can contract to narrow the arterioles which supply blood to an organ so greatly reducing the blood flow. On the other hand the muscle can relax, so opening the arterioles wide and allowing large amounts of blood to flow into a tissue. The main ways in which the behaviour of the arteriolar smooth muscle can be controlled are:

1. Sympathetic nerves which release noradrenaline (norepinephrine). These go to all the arterioles in the body and the noradrenaline which they release causes constriction of the vessels. In the 'resting' state, the sympathetic nerves are moderately active so that the vessels are neither wide open nor tightly shut. If the number of impulses passing down the nerves is increased, more noradrenaline is released and the arterioles close further. If the impulses activity is reduced, the smooth muscle can relax and the arterioles open wide (dilate).

2. Adrenaline (epinephrine) released from the adrenal medulla during exercise, fear, anger and other emergencies. Adrenaline differs from noradrenaline in that it has a dual action. It strongly constricts most of the arteriolar arterioles in the body apart from those in the muscles and the heart which it dilates. This is because while most arterioles are dominated by α receptors, those in the muscles and heart are dominated by β receptors. The overall action of adrenaline is therefore to reduce the flow of blood to organs like the skin, gut and kidneys which are not urgently needed in emergency situations and to increase blood flow to the muscles and heart which do need the extra blood.

3. Chemicals. Oxygen lack, carbon dioxide excess and excess of acid all tend to cause dilatation of arterioles, particularly in muscle, in the brain and in the heart. If the oxygen content of the blood falls, if its carbon dioxide content rises and if excess acids are being produced, these are indications that the tissues are not getting enough blood. It therefore makes sense that such chemical stimuli should dilate the arterioles in order to increase the flow of blood to a tissue.

Many other chemicals and hormones can alter the behaviour of arterioles but in most cases we do not yet understand their significance and so they will not be further considered here.

CAPILLARIES

The capillaries, which have walls only one cell thick, are the smallest vessels in the body. Only in the capillaries can exchange of materials take place between the blood and the tissues: in all other blood vessels the relatively thick walls prevent such exchange occurring. In the capillaries there is a delicate balance between the blood pressure pushing fluid out and the osmotic pressure of the plasma proteins pulling fluid back (see chapter on the body fluids). The blood pressure tends to be slightly higher than the osmotic pressure at the beginning of a capillary and so fluid tends to leave the vessel there. By the end of the capillary the pressure of the blood has fallen while the osmotic pressure of the plasma proteins is virtually unchanged: this means that the plasma protein osmotic pressure is usually higher than the blood pressure and so fluid is drawn back into the vessels. There is often a slight excess of outward over inward movement of fluid and this excess is drained away from the tissues in the form of lymph.

VENULES AND VEINS

The capillaries run together to form tiny vessels which are sometimes called venules and the venules run together to form veins. The veins have thin but relatively muscular walls and under normal conditions they contain about 60% of all the blood in the vascular system. The muscles in the vein walls are controlled by sympathetic nerve fibres and when they contract they can push a surprisingly large volume of blood out of the veins and towards the heart. On the other hand if the smooth muscle in the veins relaxes completely and the vessels dilate fully, the venous system may be capable of holding 90% of all the blood in the circulatory system.

All the veins except the very largest ones have valves. These are important because they prevent the pressure due to the whole column of venous blood being transmitted down the veins to the feet. If the valves are not functioning properly, the leg veins are subjected to a high pressure and become tortuous and dilated (varicose veins). The high pressure is also transmitted to the capillaries and excess fluid escapes from the blood in the lower part of the legs and the feet and oedema results.

The valves are particularly important during exercise. Because they do not allow the blood to flow backwards, when the veins are compressed by the muscles the blood must flow on towards the heart, thus helping the circulation. Normally in the legs the valves allow blood to flow only from the superficial veins into the deep veins of the muscles. If the valves of these so-called perforating vessels are not functioning, during exercise venous blood at high pressure is forced out of the muscles back into the superficial veins. This back flow is an important factor in the causation of varicose veins.

CIRCULATION THROUGH SPECIAL REGIONS
Many of the individual organs have peculiarities about the way in which the blood flow through them is controlled. Some of the most important organs are briefly considered in this section.

THE BRAIN
The nervous control of brain blood vessels is ineffective. Instead, the dilatation and constriction of the arterioles is determined almost entirely by the local chemical conditions. The most powerful stimulus for opening up the cerebral arterioles is an accumulation of carbon dioxide which indicates that not enough blood is reaching that part of the brain. Oxygen lack also has a strong dilating action. Blood flow through the brain thus depends almost entirely on local conditions. The brain demands and usually receives the blood flow it requires irrespective of what is happening elsewhere in the body.

THE HEART
Circulation through the heart too is controlled almost entirely by local chemical factors, although circulating adrenaline is a strong dilator. Oxygen lack is the most important stimulus, carbon dioxide accumulation being rather less important than in the brain. Again the heart demands blood, irrespective of what is happening elsewhere.

SKELETAL MUSCLES
In the muscles nervous control by sympathetic fibres, circulating adrenaline and local chemical factors are all important. During rest, the blood flow through the muscles is very low indeed. However, when exercise begins and the muscles contract, oxygen is quickly used up and carbon dioxide and acids accumulate. These cause arteriolar dilatation which is accentuated by the action of circulating adrenaline from the adrenal medulla.

THE SKIN
The skin is an organ which, at least in the short term, is not vital to life and which primarily acts as the servant of the rest of the body, particularly in the regulation of body temperature. Because of this, sympathetic nervous control of the skin arterioles is very powerful: it regulates skin blood flow in accordance with the needs of the rest of the body, often over-riding the local requirements of the skin. When the body is cold, skin blood flow is cut to the minimum to reduce the loss of heat. When the body is warm, the skin circulation is opened up in order to increase the rate of heat loss. Only when the skin is in actual danger of either freezing or burning do local factors tend to predominate: in both cases skin blood flow increases in order to reduce the risk of damage.

THE GUT
In the gut, too, nervous control by noradrenaline released by the sympathetic system is very strong. When the gut is full of food, the vessels are open wide in order to allow plenty of blood to flow in for digestion. During exercise or any other emergency, however, the arterioles constrict and gut blood flow is cut to the bare minimum, the blood being diverted to tissues where it is urgently needed. This is why it is not advisable to take vigorous exercise soon after a meal: the blood is diverted from the gut and so digestion is delayed.

THE KIDNEYS
In order to carry out their functions the kidneys require a very high blood flow for most of the day. They normally receive about 25% of the cardiac output. However, in emergency situations such as haemorrhage and exercise the blood is diverted from the kidneys to organs which are more immediately vital and so in these circumstances the flow of urine may be considerably reduced.

THE CONTROL OF BLOOD PRESSURE
If the cardiovascular system is to function satisfactorily, the pressure within the arteries must be neither too high nor too low. It must be sufficient to push blood to the top of the head and to ensure an adequate supply of blood for all organs at all times. But it must not be so high that blood vessels are damaged or burst or that the heart has to work excessively hard in order to pump out blood against the pressure in the aorta.

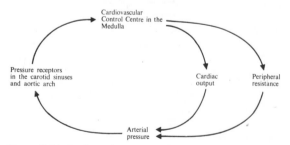

Figure 7.14 Outline of the control system which regulates arterial pressure.

As always when it is necessary to regulate something fairly precisely, the body has an effective system for regulating arterial pressure. This system, although complex in detail, is relatively simple in principle. As with all control systems, there are receptors for collecting information, there is a control centre which receives this information and decides what must be done and there are effectors which carry out the instructions of the control centre.

PRESSURE RECEPTORS OR BARORECEPTORS

The most important receptors which collect information about arterial pressure are situated in the major arteries and in particular in the arch of the aorta and the region known as the carotid sinus. The carotid sinus is a dilatation where the common carotid artery divides into internal and external branches. The receptors fire off nerve impulses when the arterial wall is stretched and the greater the degree of stretch the faster are the impulses generated. Since the higher the pressure the more the vessel wall is stretched, the patterns of impulses in the nerves from the receptors are a measure of the arterial pressure. With each systolic pressure rise, the burst of impulses gives the control centre an indication of the pressure level in the major arteries.

THE CARDIOVASCULAR CONTROL CENTRE

The main centre for the control of arterial pressure is in the medulla oblongata at the lower end of the brain. Other parts of the brain such as the hypothalamus have important parts to play but the medulla seems to be the most important region. The centre receives information from the pressure receptors about the situation in the major arteries. If pressure is normal then the control mechanism leaves well alone. If the pressure is too low the centre sends out instructions to the effectors which tend to raise pressure, while if it is too high the effectors are instructed to lower pressure.

CARDIOVASCULAR EFFECTORS

Pressure in the major arteries depends partly on how quickly blood is pumped into the aorta by the heart and partly on how quickly blood can leave the arteries via the arterioles. The faster the blood is pumped in, the higher will be the pressure: in other words the higher the cardiac output the higher the pressure will be. On the other hand, the more slowly the blood drains away via the arterioles the higher will be the pressure. If the arterioles offer a great deal of resistance to the flow of blood, the blood will drain slowly and pressure will tend to be high. If the arterioles are wide open, they will offer little resistance: blood will drain away rapidly and pressure will tend to be low. The resistance to flow offered by all the peripheral vessels and which depends mainly on the arterioles is usually termed the peripheral resistance. The blood pressure therefore depends on the balance between cardiac output and peripheral resistance. If either the cardiac output or the peripheral resistance rise pressure will tend to go up. If either falls, pressure will tend to go down. The control centre can therefore regulate arterial pressure by altering the cardiac output, the peripheral resistance or, more usually, both together. The part of the control centre which regulates the heart is sometimes known as the cardiac centre while the part which regulates peripheral resistance is known as the vasomotor centre. The main mechanisms available to the control centre are as follows:

1. Cardiac output may be altered by changing
   a. The heart rate. Noradrenaline released by sympathetic nerves speeds the heart up, acetyl choline released by the vagus slows the heart down and adrenaline released into the blood by the adrenal medulla speeds it up.
   b. The vigour with which the heart contracts. The more vigorously the heart beats, the greater is the stroke volume (the amount of blood pumped out by the heart each beat). The vigour of beating may be increased by noradrenaline released by sympathetic nerves and by adrenaline released by the adrenal medulla. An increase in venous return brought about by venous constriction will stretch the heart and also make it contract more vigorously.

2. Peripheral resistance may be altered in many ways but probably most importantly by changing
   a. The activity of sympathetic nerves to the arterioles which release noradrenaline and which can cause constriction of almost all arterioles but especially those in the gut, skin and kidneys.
   b. The rate of release of adrenaline from the adrenal medulla. The adrenaline dilates arterioles in muscle and heart but constricts those elsewhere.

COMPLICATIONS OF PRESSURE REGULATION

The outline of pressure regulation just given is true of most circumstances but there are a number of situations in which pressure in the major arteries may be normal yet nevertheless the blood flow and oxygen supply to an organ may be defective. In these circumstances, pressure in the major arteries may be raised above normal in an attempt to restore to normal levels the blood and oxygen supply to a key organ. Three organs which seem to have extra defence mechanisms which attempt to safeguard their blood supply are the brain, the kidneys and the pregnant uterus.

The Brain

If the drainage of cerebrospinal fluid is blocked in some way, the continued secretion of the fluid causes pressure inside the skull to rise. This rising pressure will tend to reduce the inflow of blood into the cranial cavity. Thus even though pressure in the carotid and vertebral arteries is normal, brain blood flow may be defective. The brain is clearly aware of this although as yet we do not understand the mechanisms involved. The result is that the control centre orders a rise in pressure in the major arteries in an effort to maintain a normal blood flow to the brain in the face of the increased resistance. Thus a rise in arterial pressure is an important clinical feature in patients whose intracranial pressure is rising.

Another factor altering blood pressure which is probably related to the need to supply adequate oxygen to the brain is the oxygen level in arterial blood. If the arterial oxygen level falls (or if the carbon dioxide level or acidity rise) receptors in the aortic and carotid bodies (chemoreceptors) fire off impulses and blood pressure rises, presumably in an effort to increase the rate at which oxygen is supplied to the brain. The aortic and carotid bodies are tiny pieces of tissue close to the pressure receptors in the aortic arch and carotid sinus.

The Kidneys

If the kidneys are to work effectively to remove waste material from the blood they must receive a large supply of blood at an adequate pressure. If they do not receive such a blood supply, waste will accumulate in the body and eventually cause death. It is not uncommon for the kidneys to be damaged by disease or for the renal arteries to be partially blocked by atheroma or by an overgrowth of the tissue of the arterial wall. In either case the effect is that the kidney cannot carry out its functions satisfactorily. In an attempt to return renal function to normal, the kidney then seems to set in motion a train of events which raises the central arterial pressure to abnormally high levels. The details remain obscure but it is possible that the main events are as follows. Cells near to the glomeruli of the kidney (see chapter 9) release an enzyme known as renin. Renin can act on a protein in the plasma to break off from it a substance known as angiotensin which is a very powerful constrictor of arterioles. As well as constricting arterioles, angiotensin can act on the adrenal cortex to release aldosterone which increases the amount of salt and water retained in the body by the kidneys. Angiotensin and aldosterone may be responsible for the pressure rise.

The Pregnant Uterus

It is obvious that the pregnant uterus must have an excellent blood supply in order to allow the development of a healthy baby. If for some reason such as narrowed uterine arteries the blood flow to the uterus is impaired, there is some evidence that the uterus initiates a series of events which raises aortic pressure in a presumed attempt to raise uterine blood flow to normal. As yet the mechanisms involved are not understood.

RESPONSES OF THE CARDIOVASCULAR SYSTEM TO SPECIAL SITUATIONS

In this section the responses of the cardiovascular system as a whole to various special situations will be discussed.

POSTURAL CHANGES

Because veins are so thin-walled and stretchable their degree of filling is very much affected by the force of gravity. This can readily be seen in a subject with prominent veins by asking him first to hold his hand above his head and then to lower it so that it hangs by his side. With the arm above the head the veins usually cannot be seen because the blood drains out of them under the influence of gravity. But with the hand by the side, gravity increases the pressure of blood in the veins and as a result they become prominent and full of blood. If someone is lying down flat the veins in the lower part of the body tend to be relatively empty of blood. On standing up gravity raises the pressure of blood in the leg veins which are therefore stretched and dilated. As a consequence of this, the filling of these dilated veins temporarily reduces the flow of blood back to the right heart. Since the veins when fully stretched can hold as much as 90% of the total blood volume and since the heart can pump out only that blood which it receives, this pooling of blood in the veins will tend to lower the cardiac output and therefore also the blood pressure. If the blood pressure falls far enough, the blood supply to the brain will be impaired and the person will become unconscious. In normal individuals this sequence of events is prevented by a reflex which operates when a person moves from the lying to the upright position. Muscle, joint and inner ear receptors send information to the control centre that the person is standing up. Immediately the centre orders an increase in activity in the sympathetic nerves which go to the veins. This makes the muscle in the vein walls contract. The veins are thus stretched less than they would otherwise be, the pooling of blood is limited and cardiac output and blood pressure are maintained. In people whose sympathetic system is defective because of disease, injury, surgical interference or drugs, the reflex is blocked and fainting is very likely to occur on suddenly standing up from a lying position. Fainting can happen in normal people who get out of a hot bath too suddenly as the warmth puts the vein walls into a very relaxed state.

EXERCISE

The cardiac output at rest is 4–5 L/min and during exercise this may rise to 30 L/min or even more. Despite this potential seven fold increase in cardiac output, arterial pressure rises by not more than 50% and there must therefore be a considerable reduction in the peripheral resistance in order to accommodate the large cardiac output change. The dilatation of arterioles takes place in the skeletal muscles and heart. Arterioles in the gut and kidneys may constrict strongly. Thus blood is diverted from those organs which are of no immediate value in exercise to those which are vital.

FAINTING

Fainting or syncope is a loss of consciousness due to a fall in pressure in the arteries supplying the brain. It may be initiated in two quite separate ways. First, any severe emotional shock can provoke a faint. Second, any persistent fall in cardiac output which tends to lead to a fall in blood pressure may lead to a faint. In both cases the effector mechanisms are the same: a great increase in vagal activity slows the heart and thus reduces cardiac output: simultaneously the blood vessels in the muscles dilate thus lowering the peripheral resistance. The net result is a sudden severe drop in blood pressure. This reduces the blood flow to the brain below a critical level and the person falls to the floor unconscious.

Paradoxical though it may seem, a faint may be an important adaptive response which aims to maintain the cerebral blood flow in difficult circumstances. It may seem ludicrous that a tendency for arterial pressure to fall should be 'treated' by the control centre inducing a sudden further fall in pressure which leads to unconsciousness. Common causes of fainting are standing still in hot conditions for long periods (as with soldiers on parade or nurses watching a surgical operation) and loss of blood. When standing still in a hot place the leg veins become dilated. In the absence of leg movements, more than 80% of the blood may become pooled in the veins in the lower part of the body. This so reduces the amount of blood effectively available to the heart that the cardiac output begins to fall and with it the blood pressure. If pressure cannot be maintained by vasoconstriction (most obviously noted by observing a dead white skin) then a faint is initiated. This rapidly forces the person to assume the horizontal position and so restores the flow of blood to the brain. A similar thing is true of haemorrhage: the horizontal position allows the heart to supply the brain most easily in conditions when the effective blood volume is reduced.

It follows from this that the worst thing anyone can do to someone who has fainted is to sit them up. The low blood pressure may then reduce the brain flow to dangerous levels and death has occurred when people have fainted in situations where they cannot fall such as a crowded train or a telephone booth. If you must do something for someone who has fainted leave him flat and lift his legs into the air to help the drainage of blood back to the heart.

HAEMORRHAGE

Bleeding or haemorrhage may be classified into three grades according to the nature of the pathophysiological response. In practice the three grades merge into one another with no sharp dividing lines.

1. Bleeding which is so slow that it causes no fall in either cardiac output or blood pressure. Examples are menstrual bleeding or bleeding from a chronic peptic ulcer. Extra water and salt are retained to compensate for the loss and the rates of synthesis of the plasma proteins and the formed elements are increased. Provided that nutrition and iron supplies are adequate a healthy individual can easily compensate for this type of haemorrhage. The commonest cause of trouble is an inadequate intake of iron with consequent iron deficiency anaemia.

2. Bleeding which causes a fall in venous return and cardiac output but no fall in arterial pressure. This may occur if a small artery or vein is damaged. The blood volume, venous return and cardiac output fall and these changes tend to make blood pressure fall too. However the tendency for pressure to fall is detected by the pressure receptors and the control centre initiates constriction of arterioles in order to compensate for the falling cardiac output. Pressure is therefore maintained at normal levels: this can lull the doctor into a sense of false security because if he does not understand the physiological mechanisms he may feel that because pressure is normal the patient is not bleeding.

The increase in peripheral resistance itself initiates changes which help to maintain blood volume. When an arteriole constricts the pressure in the capillary beyond it must fall. The osmotic pressure of the plasma proteins is unchanged and so the force drawing fluid into the capillary is greater than that pushing fluid out. There is therefore a tendency for fluid to move from the extravascular compartments into the blood. Plasma proteins and formed elements cannot be replaced in this way and so the blood becomes diluted with a fall in haematocrit and haemoglobin levels.

If the haemorrhage is arrested by the haemostatic mechanisms (chapter 6) then the body compensates by more vigorous responses of the type outlined in 1. If the haemorrhage is not halted the situation passes into stage 3.

3. Bleeding which leads to a fall in venous return, in cardiac output and in arterial pressure. The cardiac output falls to levels so low that a normal blood pressure can no longer be maintained by increasing the peripheral resistance. Eventually a faint may be initiated and the pressure falls catastrophically. The main features of this type of haemorrhage are:

a. A falling blood pressure.

b. A weak pulse because the stroke volume is low.

c. A greyish-blue skin pallor. Vasoconstriction reduces the flow of blood and the slow flow allows haemoglobin to become completely deoxygenated.

d. Dry mouth and lax skin due to movement of fluid from the extravascular compartment into the blood.

e. Rapid, shallow breathing. The low rate of blood flow through the chemical receptors in the aortic and carotid bodies stimulates breathing (chapter 8).

f. Changes in the distribution of the cardiac output. Skin, gut and kidney blood flow are cut down drastically. In severe haemorrhage the lack of an adequate blood supply to the kidneys can cause renal damage and failure.

g. The outputs of ADH and aldosterone (chapter 9) are increased so that if the kidney is still working it holds back maximal amounts of salt and water.

CARDIOVASCULAR SHOCK

Cardiovascular shock is said to occur whenever the cardiac output falls below the level required by the body. The signs and symptoms are similar to those of severe haemorrhage but they may be initiated in several different ways:

1. *Cardiogenic shock.* This occurs whenever the fall in cardiac output is primarily due to a defect in the heart such as a coronary thrombosis.

2. *Shock due to a failure of venous return.* There may be an actual loss of blood as occurs during haemorrhage or there may be a loss of fluid from the vascular system into the tissues because of damage to capillary walls as occurs in burns and crush injuries. Shock after surgery may be due partly to blood loss, partly to fluid loss into the tissues and partly to an inadequate fluid intake which prevents normal compensatory mechanisms from operating.

If blood has been lost, the only effective treatment is transfusion with whole blood. If compatible blood is not available, plasma or some plasma substitute should be used. Saline should be employed only in extreme emergency if nothing else is available because since it contains no dissolved proteins or equivalent substances most of it quickly leaves the blood and enters the extravascular tissues. With burns, since formed elements have not been lost to a great extent, plasma is the treatment of choice. In some cases in spite of all treatment the blood pressure and cardiac output fail to rise. This situation is called irreversible shock and the reason for it is unknown. It may be due to irreversible damage to the heart or to the arterioles.

8 The Respiratory System

The respiratory system is a mechanism for moving oxygen from the air to the lungs and then into the blood and for moving carbon dioxide in the opposite direction. Because carbon dioxide when dissolved in water becomes an acid, the lungs are very important in the regulation of acid-base balance.

STRUCTURE

The exchange of gases between air and blood takes place in minute blind sacs known as alveoli. Each alveolus has a wall consisting of a single layer of flattened cells in intimate contact with blood capillaries. Here the air and the

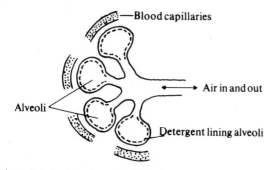

Figure 8.1 Outline of alveolar structure.

deoxygenated blood pumped into the lungs by the right side of the heart come into close contact. Since the amount of oxygen in the air is greater than that in the deoxygenated blood, oxygen automatically tends to flow down a concentration gradient from the air into the blood. The reverse happens with carbon dioxide. Thus by passing through the lungs, the venous-type blood which is poor in oxygen and rich in carbon dioxide is converted to arterial-type blood rich in oxygen and relatively poor in carbon dioxide.

The air reaches the alveoli via a system of tubes consisting of trachea, bronchi and bronchioles. These contain smooth muscle fibres and the larger ones also contain cartilage for support. The smooth muscle in the

bronchi and bronchioles is particularly interesting because its receptors are predominantly of the β type. This means that it is relaxed by adrenaline and by the β stimulating drugs thus explaining why these substances may be helpful during the respiratory distress associated with asthma.

All the tubes except those immediately preceding the alveoli which are known as alveolar ducts are lined by an epithelium consisting of mucus-secreting goblet cells and ciliated cells. The cilia on the cell surfaces beat upwards towards the trachea and the throat thus generating a constant upward stream of mucus. This traps all types of foreign particles, including bacteria, and carries them upwards and outwards thus helping to keep the lungs clean. The cilia may be paralysed by toxic gases, by industrial fumes and by cigarette smoke thus making it difficult to keep the lungs clean.

Only the alveoli themselves are concerned with gas exchange. The air in the nose, mouth, trachea, bronchi, bronchioles and alveolar ducts cannot make effective contact with the blood. Because it is not used in the actual process of gas exchange the space within the tubes leading from the nose and mouth to the alveoli is known as the 'dead space'.

THE MECHANISM OF BREATHING

The lungs are contained within the cavity of the chest. Contrary to what one might expect, the lungs are not attached to the wall of the chest by any actual tissue connections. Instead there is a remarkable arrangement which it is essential to understand if some of the diseases of the chest are to be properly appreciated.

The lungs themselves are covered by a smooth, moist membrane known as the visceral pleura. The chest wall is lined by a similar smooth, moist membrane known as the parietal pleura. Only the parietal pleura is pain-sensitive. In normal individuals these two layers of pleura are closely applied to one another. Although there are no actual bridges of tissue between them the two layers never normally come apart. When the chest wall expands, the visceral pleura is pulled outwards in company with the parietal pleura and the visceral pleura pulls out the underlying lungs. The pleural cavity is the space between the two pleural layers. Normally it contains nothing but a very thin film of liquid and it is thus more of a potential space than a real one in normal people.

The only link between the two smooth moist surfaces is a thin film of water. Because of the force of surface tension, the water molecules in this film strongly resist being pulled apart from one another and from the pleural membranes. Therefore when the parietal pleura moves outwards as the chest expands, it pulls on the water molecules which again pull on the visceral pleura. The two pleural layers are therefore effectively stuck together because of the existence of this thin water film.

Lung tissue is highly elastic and outside the body the lungs shrink to a fraction of their volume in the chest. They are therefore kept expanded when in the body only by the force exerted by the thin film of pleural fluid which keeps the visceral and parietal pleura stuck together. If either the chest wall or an alveolus is punctured, so allowing air into the pleural cavity, the lung collapses giving the condition known as pneumothorax. Fortunately the two pleural spaces are not continuous and unless both sides are damaged simultaneously, only one side will collapse at a time.

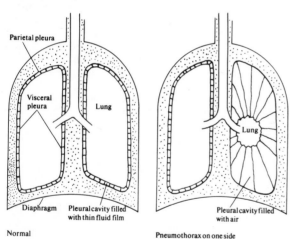

Figure 8.2 The lungs and pleural membranes showing how a pneumothorax can occur.

The main muscles of respiration are the intercostal muscles between the ribs, and the diaphragm, the large sheet of muscle which separates the chest from the abdominal cavity. Breathing in is achieved by moving the ribs upwards and outwards and the diaphragm downwards, so increasing the volume of the chest cavity. Because of the pleural mechanism the lungs are also forced to expand. Since the air within must now occupy a bigger volume its pressure falls below the pressure in the outside atmosphere. Air therefore must move in via the respiratory passages until the pressure inside the lungs is again equal to that in the atmosphere.

Breathing out is achieved by the reverse process. The ribs move down and inwards and the diaphragm rises. This reduces the volume of the chest and the air within the lungs becomes slightly compressed. This compression raises its pressure above atmospheric and the air moves out of the lungs until the pressure inside is again equal to that in the atmosphere.

During breathing in, the chest expansion pulls on the bronchi and bronchioles and opens them wider. During breathing out, the collapsing chest presses on the tubes and narrows them. This explains why in asthma where the tubes are narrowed by smooth muscle contraction, breathing out is often much more difficult than breathing in.

Coughing is a special pattern of breathing in and out which tends to expel foreign matter from the respiratory tubes. After an initial deep inspiration (breathing in), a forced expiration is made. During the expiration the glottis, the mechanism in the larynx which guards the entrance to the trachea, is initially closed so that no air can actually leave the chest. This means that pressure builds up rapidly and when the glottis suddenly opens, air rushes out at a speed which may approach 500 mph. This tends to carry solid and liquid matter up and out.

LUNG VOLUMES AND LUNG FUNCTION TESTS
Names for the various volumes of air within the respiratory system have now become widely accepted conventions. The tidal volume is the volume of air breathed

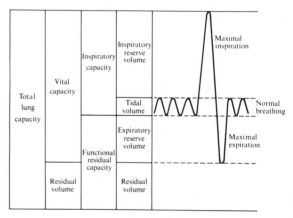

Figure 8.3 The relationships between the various volumes of air in the lungs.

in and out during one respiratory cycle. The vital capacity is the volume of air moved out of the lungs when a maximal inspiration is followed by a maximal expi-

Table 8.1 Approximate values for vital capacity and residual volume in various classes of people. Volumes are in litres (6.1 EP).

	Young males	*Young females*	*Old males*
Vital capacity	4·8	3·2	3·6
Residual volume	1·2	1·0	2·4
Total lung volume	6·0	4·2	6·0

ration. The residual volume is the amount of air left in the lungs after a maximal expiration. Some approximate normal values are shown in table 8.1.

Many attempts have been made to develop reliable tests for measuring the state of lung function in patients but most have proved either unreliable or suitable only for research purposes. Three may be used routinely:

1. *Vital capacity.* The patient is asked to take a deep breath in and then to breathe out again into an instrument which measures the volume and records it on a chart.

2. *Forced vital capacity (FVC) or forced expiratory volume (FEV).* In patients with narrowed bronchi and bronchioles the total vital capacity may be normal but the time taken to blow out the air may be greatly prolonged. In a normal individual about 80% of the air is expelled in one second. The patient is therefore asked to take a deep breath in and then to breathe out as quickly as he can into a recording instrument. The volume of air breathed out in one second is recorded.

3. *Peak flow.* This is a rather simpler way of noting how quickly a patient can blow out air. He is asked to take a deep breath and then to blow out as quickly as possible into a device known as a peak flow meter which records the peak rate at which the air passes through it.

THE WORK OF BREATHING
The work which must be expended in order to breathe is used in overcoming two different types of resistance:

1. *Friction.* When air moves along the bronchi and bronchioles there is a frictional force between the moving air and the tube wall. The faster the rate at which the air moves and the narrower the tube the greater is the frictional resistance in relation to the volume of air which is moved.

2. *Elasticity.* The lungs and chest wall are both elastic. At the end of a quiet expiration the elastic forces balance out and no movement of the lungs and chest wall would occur if death suddenly occurred at this point. But if the lungs and chest are expanded or contracted beyond this point an elastic force must be overcome. As when stretching an elastic band, the effort required to move the chest increases the further one moves from the equilibrium position. Thus when breathing is slow and deep a great deal of effort is required to overcome the elastic forces.

The frictional force which must be overcome is greatest when breathing is rapid and shallow, while the elastic force is greatest when breathing is slow and deep. Obviously there must be some optimum intermediate rate of breathing when the sum of the two forces which must be overcome is at a minimum. In practice it has been found that the respiratory control centre is a remarkable computer which acts to minimise the amount of work which must be done during breathing. At any given ventilation rate (the ventilation rate is the respiratory equivalent of the cardiac output and is the volume of air moved in and out of the lungs in one minute) the brain adjusts the rate and depth of breathing so that the sum of the work done in overcoming friction and that done in overcoming elastic recoil is at a minimum. The control centre can perform this remarkable feat only because it receives a constant barrage of information from joint and muscle receptors in the chest wall and stretch receptors in the lung tubes.

SURFACE TENSION AND THE LUNGS
Some of the work done in overcoming the elastic recoil of the lungs is used to stretch the solid tissue but some is used in stretching the thin layer of fluid which lines every alveolus. Inflating a lung is rather like inflating a million miniature soap bubbles. As almost anyone who has played with soap knows, when you blow a soap bubble you must keep blowing if the bubble is to continue expanding. If you stop blowing the bubble immediately starts to collapse. This is because of the force of surface tension. Molecules in a thin liquid film resist being pulled apart: when an active force pulling them apart ceases they rush together again.

The force of surface tension may be greatly reduced by chemicals known as detergents which reduce the attraction of water molecules for one another. In the presence of a detergent the effort which must be expended in order to pull the molecules apart is much less. It is therefore not surprising that the body has evolved a mechanism whereby detergent is secreted into the alveoli. This reduces the surface tension of the thin fluid film to about one tenth of what it would otherwise be and greatly reduces the work of breathing.

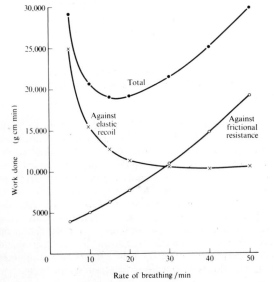

Figure 8.4 The relationships between the amount of work done in overcoming elasticity and the amount done in overcoming friction at various frequencies of breathing.

The effectiveness of the detergent (often known as 'lung surfactant') is best seen by noting what happens when the detergent is absent. This occurs in some babies who are born prematurely. They are unable to manufacture the surfactant and they develop a condition known as the respiratory distress syndrome in which it is obvious that the baby is having to work incredibly hard in order to breathe. This is not surprising because in the absence of surfactant it is necessary to work about ten times as hard as normal in order to overcome the surface tension.

THE CONCEPT OF PARTIAL PRESSURE

Air is made up of about 21% of oxygen and 79% of nitrogen. Other gases such as carbon dioxide and the 'noble gases' like argon and xenon make up well under 1% of the air.

This means that of the column of air stretching up from the earth's surface, 21% consists of oxygen. Therefore 21% of the total pressure which that column of air exerts is due to oxygen. The total air pressure at sea level is in the region of 760 mmHg. 21% of 760 is 160 and so we say that the 'partial pressure' of oxygen in air at sea level is 160 mmHg. It is called the partial pressure because it is that part of the total atmospheric pressure which is due to oxygen. The partial pressure of nitrogen at sea level is therefore about 600 mmHg. The partial pressure of a gas is normally symbolised by a small p followed by the chemical symbol for the gas (e.g. pO_2, pCO_2, pN_2).

As one goes up to higher altitudes, the height of the column of air above decreases and the total atmospheric pressure falls. Suppose that one goes up in an aircraft to a height where the total atmospheric pressure is just half that at sea level, namely 380 mmHg instead of 760 mmHg. At that altitude, 21% of the air is still oxygen and 79% is still nitrogen. But the partial pressure of oxygen is 21% of 380 which is 80 mmHg and the partial pressure of nitrogen is 79% of 380 which is 300 mmHg. The percentage composition of the air remains the same but the partial pressures fall with increasing altitude.

When a liquid is shaken up in the presence of a gas, some of the gas dissolves in the liquid. When the two have come into equilibrium, we say that the partial pressure of the gas in the liquid is the same as the partial pressure of the gas which is in contact with the liquid. Thus, if blood is shaken up with air, oxygen enters the blood until the partial pressure of oxygen in the blood is 160 mmHg. Nitrogen enters until its partial pressure is 600 mmHg.

CONTROL OF VENTILATION RATE

The ventilation rate can be defined as the volume of air moved in and out of the lungs every minute. The ventilation rate which is required depends to a large extent on the amount of muscular exercise being performed although of course even in the resting state some oxygen is required and some carbon dioxide is produced. But during exercise much more oxygen is used up and much more carbon dioxide is produced, so that if normal blood

levels of these gases are to be maintained the ventilation rate must increase accordingly.

How does the body know what the ventilation rate should be? There is a typical control system with sensory receptors, a control centre and effectors. The control centre is in the lower parts of the brain known as the medulla and pons and it is usually called the respiratory centre. The effectors are the muscles of the chest wall,

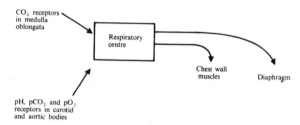

Figure 8.5 The respiratory control system.

mainly supplied by the thoracic spinal nerves, and the diaphragm, supplied by the phrenic nerve which leaves the spinal cord in the neck and travels down through the thorax. The receptors are in several different places. The most important ones monitor the chemical composition of arterial blood. Normally the partial pressure of oxygen (pO_2) is 95–100 mmHg, that of carbon dioxide (pCO_2) is about 40 mmHg and the pH is about 7·4.

THE RECEPTORS

The most important ones are as follows:

1. Receptors in the brain near the respiratory centre. These are well placed to monitor the arterial blood passing through the brain and also possibly the cerebrospinal fluid. They are sensitive to carbon dioxide. A rise in pCO_2 stimulates ventilation while a fall in pCO_2 reduces ventilation. The brain receptors are not sensitive to oxygen lack in that oxygen lack does not stimulate ventilation by acting on them. A deficiency of oxygen at the brain level in fact depresses ventilation because it interferes with the functioning of nerve cells in the respiratory centre.

2. Receptors in the carotid body and the aortic body. These are small pieces of tissue near the carotid sinus and in the wall of the arch of the aorta: they lie close to the corresponding pressure receptors. They are sensitive to three separate chemical stimuli, the pH and the levels of oxygen and carbon dioxide in arterial blood. A rise in acidity, a fall in pO_2 and a rise in pCO_2 all stimulate the receptors and lead to a rise in ventilation rate. The reverse changes lower the ventilation rate. By acting on these receptors (sometimes known as the peripheral chemoreceptors) oxygen lack therefore stimulates breathing: this is in contrast to its action on the brain.

3. Receptors in the joints which signal when joint movements occur. These warn the respiratory centre that

exercise is beginning and are responsible for an initial increase in ventilation which occurs in anticipation of the later changes in carbon dioxide and oxygen levels.

THE RESPIRATORY CONTROL CENTRE

The information from all the receptors is sent to the respiratory centre which can thus build up a clear picture of what is happening. If carbon dioxide is accumulating in arterial blood, this means that the lungs are not working rapidly enough to get rid of all the carbon dioxide that is being produced in the tissues. If oxygen levels are low in the arteries, this means that oxygen is being used up more rapidly than it is being supplied by the lungs. If acids other than carbon dioxide are accumulating in arterial blood, this means that something is wrong in the tissues: abnormal amounts of acid may be being produced because of oxygen lack or insulin lack or acids may not be being excreted normally because of kidney failure.

The respiratory centre continually adjusts the ventilation rate in order to maintain the composition of the arterial blood normal and constant. Under most circumstances it seems to be most sensitive to changes in pCO_2, probably because quite small changes in the level of carbon dioxide can seriously upset acid-base balance. Only when oxygen lack is severe or when abnormal acids are present in high concentration do these stimuli become more important than the carbon dioxide level.

Once the respiratory centre has decided what ventilation rate is required, it then has to estimate what combination of rate and depth of breathing will achieve this with the minimum of expenditure and effort. In order to do this it relies on information from muscle and joint receptors in the chest wall and from receptors in the lungs. The end result is usually very appropriate as outlined in the section on 'The Work of Breathing'. Perhaps the most important sensory information from the lungs is carried by fibres in the vagus nerve. If the vagus nerve is sectioned, breathing immediately becomes slower and deeper although the total ventilation rate remains the same. This means that a less efficient combination of rate and depth of breathing is being used. The vagus carries information from stretch receptors in the respiratory passages and lungs which indicate the degree of filling or emptying of the lungs. As breathing in proceeds, the respiratory centre uses this information to cut off inspiration at the most appropriate time. A similar process occurs during expiration. In the absence of the vagus, the respiratory control centre has to rely on less effective information from the chest wall and breathing becomes slower, deeper and less efficient. These responses mediated by the vagus are sometimes known as Hering-Breuer reflexes after the scientists who first described them.

THE TRANSPORT OF OXYGEN

Oxygen is not very soluble in water. At the normal pO_2 in the lung alveoli and in arterial blood (about 100 mmHg) only about 3 ml of dissolved oxygen can be carried by each litre of blood. This is clearly insufficient to

supply the needs of the body for even at rest an adult about 250 ml of oxygen per minute. The problem solved because the red cells contain a red pigment known as haemoglobin. This has the remarkable property of easily and reversibly combining with oxygen, one molecule of haemoglobin (Hb) picking up four of oxygen. When the haemoglobin concentration of the blood is normal (around 140–150 g/L) and when each Hb molecule is carrying four molecules of oxygen, the red cells in one litre of blood carry about 200 ml of oxygen, over sixty times the amount which can be carried in dissolved form.

Haemoglobin which is fully oxygenated is bright red while Hb without oxygen is purplish in colour. When a patient appears bluish because his blood vessels are filled with deoxygenated blood, he is said to be cyanosed.

There are two main forms of cyanosis:

1. *Peripheral cyanosis.* When the rate of blood flow to the skin is reduced because of cold or haemorrhage, the blood that remains in the skin may become deoxygenated and blue because it is stagnant. However, in these circumstances the blood in the central arteries is usually normally oxygenated as can be seen by looking at the tongue. The small blood vessels of the warm tongue are always wide open and so the colour of the tongue reflects the colour of arterial blood.

2. *Central cyanosis.* With this condition both the tongue and the skin are blue because the cyanosis is due to inadequate oxygenation of arterial blood and indicates serious failure of either lung function or heart function.

HAEMOGLOBIN DISSOCIATION CURVE

The relationship between the pO_2 of the blood and the amount of oxygen carried by Hb is an unusual one. It

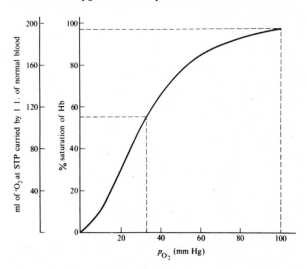

Figure 8.6 The relationship between the pO_2, the percentage saturation of haemoglobin and the amount of oxygen carried by one litre of blood.

can be described by the dissociation curve shown in fig. 8.6. At very low partial pressures below about 20 mmHg, Hb does not attract oxygen very strongly and little oxygen combines with it. But between 20 and 60 mmHg, Hb attracts oxygen very strongly indeed and the amount of oxygen bound to it rises very rapidly. At 60 mmHg the Hb is carrying about 75–80% of all the oxygen it is capable of transporting. Above this point as the pO_2 rises the amount of oxygen carried by Hb rises relatively slowly again. By the time a pO_2 of 100 mmHg is reached the Hb is more than 95% saturated. The importance of the shape of the curve lies in the following points:

1. At normal ventilation rates, with a normal alveolar pO_2 of about 100 mmHg, the Hb is almost completely saturated and each litre of normal blood carries about 200 ml of oxygen. Since the resting cardiac output is about 5 L/min, the total amount of oxygen pumped out by the heart at rest is around 1 L/min. Since the resting oxygen consumption is about 250 ml/min there is an ample safety margin.

2. The upper end of the curve is flat. This means that arterial pO_2 must fall below 60 mmHg before the transport of oxygen is seriously impaired. Even at this partial pressure each litre of normal blood carries 160 ml of oxygen and 800 ml of oxygen per minute are transported around the body so that the supply is still ample at rest.

3. Below a pO_2 of 60 mmHg the curve is steep and Hb gives up its oxygen readily and rapidly. This means that the oxygen is made available to the tissues while the pO_2 is still relatively high.

The importance of the last point is clearly seen in carbon monoxide poisoning. Carbon monoxide is an extremely toxic substance found in coal gas and vehicle exhaust fumes. Natural gas usually contains little or none of it. Carbon monoxide is so toxic because it conbines with Hb much more readily even than oxygen does and it

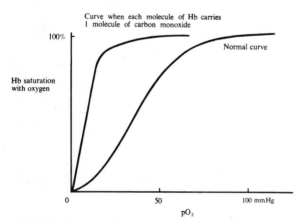

Figure 8.7 The effect of carbon monoxide on the relationship between oxygen and haemoglobin.

therefore prevents the oxygen becoming attached. However carbon monoxide can kill even when only 25% of the Hb combining points are occupied by it leaving 75% still carring oxygen. One might have thought that this would have left an ample safety margin but the real trouble arises because the carbon monoxide alters the shape of the dissociation curve. When carbon monoxide has combined with one of the four slots on an Hb molecule, the other three take up oxygen much more readily than usual and even at a pO_2 of only 20 mmHg the Hb may be 80% saturated. Looking at it the other way, the new shape of the curve means that the Hb will cling onto most of its oxygen until the blood pO_2 has fallen to 10–15 mmHg. At this blood pO_2 brain cells will be dead. Therefore, even though there may be plenty of oxygen in the blood it is useless because it cannot be given up and supplied to the tissues at a high enough pO_2.

HYPERBARIC OXYGEN
This term literally means oxygen at high pressure. If a patient is put onto a special chamber where he can breathe pure oxygen at three atmospheres pressure, the partial pressure in the lungs and arterial blood can be raised to about 2200 mmHg. At this pO_2 between 60 and 70 ml of oxygen can be carried in simple dissolved form by each litre of blood so by-passing the need for haemoglobin. The main uses of hyperbaric oxygen are:

1. Carbon monoxide poisoning. The high pressure oxygen avoids the need for Hb transport and also tends to displace the carbon monoxide from its combination with Hb.

2. Respiratory distress syndrome in infants.

3. After coronary thrombosis when the damaged heart cannot pump blood effectively through the lungs, it cannot itself receive an adequate supply of oxygenated blood and thus enters a vicious circle which can end only in death. It has been claimed that high pressure oxygen can improve the outlook although this has been disputed.

4. In the treatment of infections with bacteria which are killed by oxygen.

5. During the treatment of cancer by X-rays. It has been found that well-oxygenated tumours are more susceptible to the actions of X-rays and that the results of treatment are improved if a patient is exposed to high pressure oxygen while the cancer is being irradicated.

TRANSPORT OF CARBON DIOXIDE
When carbon dioxide enters the capillary blood from the tissues, small amounts stay in the plasma either in dissolved form or in combination with plasma proteins but most of it passes straight into the red cells. The red cells are therefore almost as important in the transport of

carbon dioxide as they are in the transport of oxygen. In the red cells two main things may happen to the carbon dioxide.

1. It may combine with Hb to form a substance known as carbamino-Hb. About 25% of the carbon dioxide which enters the blood in the capillaries is carried to the lungs in this form.

2. It may react with water to give carbonic acid and the carbonic acid may then split up to give H^+ and HCO_3^- ions.

$$CO_2 + H_2O \rightleftharpoons H_2CO_3 \rightleftharpoons H^+ + HCO_3^-$$

In the red cells there is an enzyme known as carbonic anhydrase which speeds up by about 5000 times the combination of carbon dioxide and water. The carbonic acid breaks down to give hydrogen and bicarbonate ions because these two ions are continually being removed from the red cell. This is because Hb is a very effective buffer and combines with and neutralises many of the H^+ ions as follows:

$$Hb^- + H^+ \rightleftharpoons Hb.H$$

One positive and one negative charge disappear so that the red cell remains neutral. However as the reaction moves to the right, the concentration of bicarbonate builds up rapidly and soon rises well above the bicarbonate concentration in the plasma. Because of this concentration gradient bicarbonate ions diffuse out of the red cell into the plasma. The loss of negative ions leaves the inside of the red cell relatively positive and so chloride ions diffuse in from the plasma in compensation. This is sometimes known as the chloride shift. All these changes

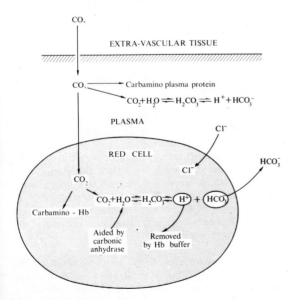

Figure 8.8 The disposal of the carbon dioxide which enters the blood in the capillaries.

are reversed when the blood gets to the lungs and the carbon dioxide diffuses out of the red cells into the alveoli where the pCO_2 is relatively very low. About 75% of the carbon dioxide which enters the tissue capillaries is carried to the lungs in the form of bicarbonate.

The red cells therefore play a potentially important role in acid-base balance and this is discussed further in chapter 12.

THE PULMONARY CIRCULATION

The pulmonary circuit from the right ventricle to the left atrium is much smaller than the systemic circuit of the circulation from left ventricle to right atrium. The pressures required are much less than those in the systemic circuit and pulmonary artery pressures are in the region of 8–15 mmHg as compared with 70–120 mmHg in the normal aorta. An important consequence of this is that capillary blood pressure in the lungs is very low. In fact it is much lower than the plasma protein osmotic pressure and so the force drawing fluid out of the alveoli into the blood is much greater than the force pushing fluid in the opposite direction. Thus the alveoli tend to be kept fairly dry and are lined by only a very thin layer of fluid which is probably actively secreted in order to keep the cells in good condition. Because the alveoli are kept relatively dry, the barrier to oxygen flow from air to blood and to carbon dioxide flow in the reverse direction is kept to a minimum.

If for any reason (such as a coronary thrombosis) the left ventricle fails, blood may pile up in the lung capillaries. Its pressure may rise above the plasma protein osmotic pressure and fluid may move out into the alveoli giving the condition known as pulmonary oedema. In the mildest form, the 'wet' lung seriously interferes with gas exchange. In the severe form the alveoli, bronchi and bronchioles may quickly be filled out with pink frothy fluid which comes up the trachea and out of the nostrils. Unless quickly treated, the acute form of severe pulmonary oedema soon ends in death.

VENTILATION AND BLOOD FLOW

In an ideal situation the distribution of both blood and air in the lungs would be even, i.e. all areas would receive similar amounts of blood and air flow per unit volume and each alveolus would receive just sufficient air to oxygenate fully the blood flowing past it. In practice this ideal is rarely achieved and many alveoli receive either too much blood for their air supply or too much air for their blood supply. As a result the blood leaving some alveoli is not fully oxygenated. While the blood leaving an alveolus where blood and air flow are in perfect balance might have a pO_2 of about 100 mmHg, the uneven distribution of blood and air means that the pO_2 of the mixed blood leaving the lungs is usually a few mmHg less than this.

The most important factor producing unevenness of blood and air flow is gravity. Changes of posture do not much affect the distribution of air but in the upright position the lower parts of the lung receive relatively more blood than the upper parts. Many forms of disease such as those caused by infections, tumours and clots also cause local maldistribution of blood and air.

There are several mechanisms available which assist in the maintenance of a balanced distribution of ventilation and blood flow.

1. If the blood supply to a region is cut off as, for example, when a clot lodges in a pulmonary artery branch, the air supply to that area of lung is reduced by the following mechanisms:

a. Secretion of the lung surfactant rapidly diminishes. This leads to a rise in surface tension with collapse of the affected alveoli.

b. Because the affected alveoli are relatively overventilated, the pCO_2 in them falls. This leads to a constriction of the small bronchioles with a diminution of the air supply.

2. If a bronchiole is blocked off by infection or by a tumour or some other factor, the blood supply to the useless alveoli beyond the block is reduced. This is achieved because the pCO_2 rises and the pO_2 falls in the affected area. Both a rise in pCO_2 and a fall in pO_2 constrict lung arterioles (in complete contrast to their effects on all other arterioles) and so the blood supply is diminished.

ABNORMAL PATTERNS OF BREATHING

The following abnormal patterns of breathing may be important in clinical medicine:

APNOEA

This term indicates the total absence of breathing. Apnoea may be brought about by an overdose of respiratory depressant drugs such as barbiturates or morphine, or by any form of damage to the respiratory control centre of the medulla.

HYPERVENTILATION

This term refers to a rate of ventilation which is in excess of the individual's need. It may occur voluntarily or may involuntarily accompany any highly emotional or excited state. It may occur during artificial respiration or be caused by some disease of the nervous system. Some drugs, notably aspirin, can cause hyperventilation by directly stimulating the respiratory control centre.

During hyperventilation the arterial pCO_2 falls and the pH rises. The high pH causes calcium ions to be bound by plasma proteins and as a result the free ionic calcium concentration of the plasma falls. This leads to increased excitability of nerve axons. In the early stages this causes firing of action potentials in response to mild mechanical stimuli. The irritation of the ulner nerve around the elbow leads to nerve impulses which cause the 4th and 5th fingers to go into spasm. Tapping on the facial nerve may cause the muscles on that side of the face to go into spasm (Chvostek's sign). Similar impulses originating in sensory fibres cause tingling sensations. With greater falls of the ionic calcium level impulses may originate spon-

taneously and there is widespread tingling and muscle twitching. The condition is known as tetany. The situation can be quickly reversed by making the patient breathe in an out of a closed receptacle such as a paper bag. The pCO_2 then rises and the symptoms rapidly and, to the patient, magically disappear.

ANAESTHESIA

The link between oxygen lack receptors and respiration is much more resistant to anaesthesia than is the link between carbon dioxide receptors and respiration. As anaesthesia deepens, first the response to carbon dioxide is lost while the oxygen lack response remains, then the response to oxygen lack is also lost and spontaneous ventilation ceases. In practice anaesthesia usually reaches only the first of these levels. The ventilation rate required to maintain an acceptable pO_2 is less than that required to maintain an acceptable pCO_2. As a result in a spontaneously breathing anaesthetised individual, the ventilation rate falls and the pCO_2 rises. At this stage, of course, since respiration is being maintained by oxygen lack, the administration of pure oxygen may cause spontaneous breathing to cease altogether with a rise in the pCO_2 to dangerously high levels.

Characteristically, when respiration is being driven primarily by oxygen lack, periods of apnoea alternate with periods of breathing. This pattern is known as Cheyne-Stokes respiration. It seems to occur because the respiratory control mechanism is relatively insensitive to small changes in pO_2. Several breaths raise the oxygen level so that the chemoreceptors are no longer stimulated. Breathing then stops until the oxygen levels fall to a point at which the chemoreceptors are again stimulated. A few breaths bring oxygen levels up again, breathing stops again and so on. This pattern of breathing is seen in a number of situations and several other complex factors may be involved in addition to the one just outlined. The main situations in which Cheyne-Stokes breathing occurs are:

1. In anaesthetised individuals who are allowed to breathe spontaneously.

2. In normal individuals in the period following recovery from hyperventilation. The hyperventilation brings the pCO_2 down to very low levels and stops breathing. However the arterial *oxygen* content is not significantly altered by hyperventilation (see earlier this chapter) and so when breathing stops, it is oxygen lack which stimulates respiration first. A few breaths restore arterial pO_2 to normal and if pCO_2 has not reached normal levels, respiration will then stop again. The cycle continues until pCO_2 levels return to normal.

3. In normal people who ascend to high altitude. In this situation the low pO_2 in the atmosphere means that oxygen lack is the primary drive to respiration. In order to maintain a normal oxygen supply, breathing must be more rapid than is necessary to get rid of carbon dioxide

at the usual rate. The carbon dioxide level therefore falls leaving oxygen lack as the only stimulus to ventilation.

4. During sleep the response to carbon dioxide is reduced and in some individuals, especially those who are elderly, the reduction may be sufficient to leave oxygen lack as the main stimulus to ventilation. Cheyne-Stokes breathing may then occur.

5. During cardio-respiratory disease in which the reduction in arterial pO_2 and rise in pCO_2 affect the behaviour of the respiratory control centre so that it is driven primarily by oxygen lack. The administration of pure oxygen in this situation may make breathing stop altogether or be greatly reduced in rate and the patient may become unconscious because of the anaesthetic effect of the resulting very high carbon dioxide levels.

BREATHING OXYGEN

In those whose lungs are malfunctioning for some reason or whose ventilation rate is inadequate, the oxygen content of the inspired gas may be raised in an effort to raise the pO_2 of arterial blood. Since simply giving oxygen cannot increase the ventilation rate, the treatment cannot increase the rate of elimination of carbon dioxide. The administration of oxygen is not without its dangers, the main ones being:

1. In new born infants, the breathing of pure oxygen causes overgrowth of retinal blood vessels and blindness, a condition known as retrolental fibroplasia. For this reason only in exceptional circumstances are new born infants given pure oxygen to breathe. A mixture containing 60% of oxygen seems to be safe.

2. Even in adults the breathing of pure oxygen for more than 24 hours can be dangerous. There seems to be interference with brain metabolism, possibly because of oxidation of the important coenzyme, lipoic acid and also, for unknown reasons exudates tend to form in the lungs. Those requiring prolonged oxygen therapy are given mixtures containing less than 60% oxygen.

3. In patients with chronic cardio-respiratory disease whose breathing is kept going by oxygen lack, administration of a gas mixture containing too high a percentage of oxygen may depress respiration and allow carbon dioxide levels to rise to a point where unconsciousness occurs.

9 The Kidneys and Urinary Tract

The main functions of the kidneys are the regulation of the amount of water and dissolved ions in the body, the excretion of waste matter and the manufacture of erythropoietin, the hormone which regulates red cell synthesis.

STRUCTURE AND FUNCTION

OUTLINE OF STRUCTURE

The principal features of kidney structure can be seen in a longitudinal section through the organ. There is a dark outer cortex with a paler inner medulla. In the centre there is a cavity, the renal pelvis, into which the urine drains. The urine is carried from the pelvis to the bladder by the ureters.

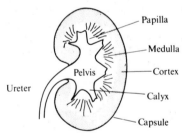

Figure 9.1 A section through a kidney showing the main regions.

The basic units out of which the kidneys are constructed are the nephrons. Each human kidney has about a million of them. Each nephron is a blind-ended tube made of a single layer of cells. It follows a tortuous route from the cortex where the first process of urine formation takes place to the medulla where the urine finally enters the pelvis. The blind end of each nephron is a cup-like structure known as Bowman's capsule. From this cup there leads a highly convoluted tube known as the proximal tubule which is still within the cortex. The tubule then takes a surprising course, dipping right down into the medulla and coming back again in a U-shaped structure

known as the loop of Henle. There then follows another tortuous section known as the distal tubule. Finally the distal tubules join together to form collecting ducts which run down through the medulla to drain into the renal pelvis.

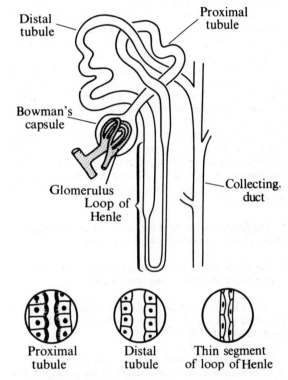

Figure 9.2 A nephron.

The details of kidney blood supply are very important. The arteries divide to give arterioles which enter the cup of the Bowman's capsule (afferent arterioles). In the capsule they break up into a mass of capillaries known as the glomerulus. The capillaries then join up to form a second arteriole (efferent arteriole) which leaves the capsule. Associated with the efferent arteriole are a group of cells known as the juxta-glomerular apparatus which make the enzyme renin.

Once the efferent arterioles leave the glomeruli they do not send the blood into veins on its way to the heart. This is why they are called arterioles rather than venules. Instead they break up into a second set of capillaries which supply the proximal and distal tubules, the loops of Henle and the collecting ducts. This second set of capillaries then join up to form venules and veins.

The kidney has a rich sympathetic nerve supply. The fibres go to two main parts of the organ.

1. The arterioles, especially the afferent arterioles. These sympathetic fibres which release noradrenaline are important in adjusting renal blood flow and especially in reducing blood flow during exercise and haemorrhage when the blood is diverted to other organs.

2. The juxta-glomerular apparatus where the sympathetic impulses can alter the amount of renin produced.

GLOMERULAR FILTRATION
Urine is initially formed in the glomeruli and Bowman's capsules. Because the renal arteries and afferent arterioles are short and relatively wide, pressure in the glomerular capillaries is considerably higher than pressure in any other capillaries in the body, being 60–70 mmHg instead of the usual 20–30 mmHg. This means that the capillary blood pressure is considerably higher than the osmotic pressure of the plasma proteins. There is therefore a tendency for all substances which can pass through the glomerular capillary wall to do so rapidly. This means that all the blood constituents apart from the plasma proteins, the platelets and the red and white cells pass into the Bowman's capsule. The important things to note about the normal glomerular filtrate, as it is called, are the following:

1. The formed elements of the blood are absent. Red and white cells do not normally appear in urine.

2. Protein is absent and therefore protein in the urine is always an indication of some abnormality.

3. Apart from the absence of protein, platelets and red and white cells, the chemical composition of the filtrate is identical to that of the plasma.

The blood flow to both human kidneys is in the region of 1200 ml/min in the resting state and the plasma flow is therefore about 700 ml/min. About one fifth of the plasma is filtered off in the glomeruli and so the glomerular filtration rate, the amount of fluid filtered per minute, is in the region of 120–140 ml/min in a normal adult. This means that each day 150–180 *litres* of fluid enter the nephrons. The normal daily urine output is less

than 2 litres and so 99% of the fluid which enters the nephrons must be salvaged and returned to the blood.

THE PROXIMAL TUBULE
The greater part of the salvage operation takes place in the proximal tubule. The most important substances which are reabsorbed from the tubular fluid and returned to the blood are the following:

1. *Sodium.* There is an active pumping mechanism which removes sodium ions from the tubules and transfers them to the blood.

2. *Chloride.* When each positive sodium ion is reabsorbed, an excess of negative charge is left in the urine while there is a relative excess of positive charge in the tubular cells. Since like charges repel while unlike charges attract, the electrical balance is kept partly because chloride ions follow the sodium.

3. *Bicarbonate* ions are reabsorbed by a complex mechanism which is partly dependent on hydrogen ion secretion into the urine and partly dependent on sodium reabsorption from the urine into the blood. Within the tubular cells carbon dioxide is produced by metabolism. Its combination with water to form carbonic acid is facilitated by the enzyme carbonic anhydrase. As sodium ions are pumped out of the urine, hydrogen ions pass from the tubular cells into the urine in part exchange. This pulls the reaction in the cells to the right and the bicarbonate which is formed diffuses out into the blood. The hydrogen ions entering the urine meet bicarbonate ions which have been filtered at the glomerulus. Since there is a steady supply of both hydrogen and bicarbonate ions, the reaction in the urine is pushed to the left with the formation of carbon dioxide and water. This raises the pCO_2 of the urine above that of the tubular cells and carbon dioxide therefore diffuses out of the urine. The whole cycle is shown in fig. 9.3.

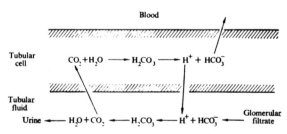

Figure 9.3 The renal tubular mechanisms for the excretion of hydrogen ions and the reabsorption of bicarbonate.

4. *Potassium* is actively reabsorbed.

5. *Glucose* is actively reabsorbed by a special pumping mechanism. Normally all the glucose is removed from the tubular fluid by the end of the proximal tubule and none appears in the final urine. However, the pumping

mechanism has a limited capacity and if too much glucose is delivered to it, it cannot cope and some of the glucose passes straight through and out into the urine. This does not normally occur until the concentration of glucose in the plasma (and in the glomerular filtrate, for the two have the same composition) rises above about 180 mg/100 ml.

6. *Water.* The reabsorption of water depends on two main factors:

a. The protein osmotic pressure of the blood in the capillaries which supply the proximal tubule. This is raised above normal because in the glomeruli about one fifth of the plasma is lost from the blood but none of the protein is lost. The protein concentration is therefore increased, so raising the osmotic pressure and the force drawing fluid from the proximal tubule.

b. The continual removal of sodium, chloride, bicarbonate, glucose and other dissolved materials from the tubular fluid means that there is a tendency for their concentrations to be slightly higher in the blood than in the tubular fluid. In order to maintain the balance, because of osmotic forces, water follows the dissolved material. This second factor is probably considerably more important than the protein osmotic pressure in water reabsorption.

All the glucose, all the potassium and most of the bicarbonate are normally removed from the tubular fluid by the end of the proximal tubule. About 80–90% of the water and sodium are also removed leaving relatively small quantities to be dealt with by the rest of the nephron.

THE LOOP OF HENLE

The functions of the loop of Henle are understood in broad outline but the details remain obscure. The loops of the nephrons form what is known as a 'counter current multiplying system'. As a result of this, the medulla becomes loaded with sodium and its accompanying negative ions. In the kidney cortex the total concentration of ions in the tubules, in the cells, in the interstitial fluid and in the plasma is similar to that in any other tissue in the body. But on moving down into the medulla, all these fluids become progressively more concentrated until in the deepest areas of the medulla the concentration may be 3–4 times that of normal plasma. The concentrated gradient is maintained because the fluid going around the loop of Henle always leaves a little of its sodium and negative ions behind, without losing any water. The fluid leaving the loop of Henle and entering the distal tubule is thus slightly more dilute than that entering the loop of Henle from the proximal tubule. The detailed counter current mechanism will not be discussed here but its significance will be explained in the section on hormonal control of the kidney.

DISTAL TUBULE AND COLLECTING DUCT

As far as dissolved substances are concerned, the main thing which happens in the distal tubule is the reabsorption from the urine of sodium in exchange for the secretion into the urine of hydrogen and potassium ions. Either hydrogen or potassium can be excreted in exchange for sodium: this means that if the hydrogen ion

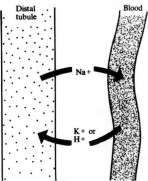

Figure 9.4 Ionic movements in the distal tubule.

secretion rate falls the potassium secretion rate must rise and vice versa. Some of the consequences which may occur as a result of this mechanism are:

1. If the blood becomes alkaline, few hydrogen ions are secreted by the distal tubule. This may lead to the loss of considerable quantities of potassium instead.

2. If the blood becomes acid, many hydrogen ions are secreted. Potassium then tends to be retained in the body.

3. If the body is severely depleted of potassium ions, hydrogen ions must be excreted by the kidneys in exchange for sodium. This hydrogen ion secretion occurs even in the presence of alkaline blood, leading to the paradoxical situation of an alkaline blood and an acid urine.

The distal tubule and collecting duct are very important in the regulation of water excretion. Their role in this will be fully discussed in the section of hormonal control of the kidney.

ACID EXCRETION

Hydrogen ions are secreted into the urine by both the proximal and the distal tubules. More are secreted by animals on a meat diet than by those on a vegetarian diet. The lowest pH which the urine can reach is about 4.5. In order to buffer the hydrogen ions which are secreted into the urine three separate mechanisms are available:

1. The hydrogen ions may combine with bicarbonate to give carbonic acid, the latter then splitting up to give water and carbon dioxide. The carbon dioxide diffuses back from the urine into the blood and is excreted by the lungs.

2. The hydrogen ions may combine with phosphate. Most of the phosphate in plasma and in the glomerular filtrate is in the form of HPO_4^- ions. Each one of these can combine with and remove from solution one hydrogen ion to give $H_2PO_4^-$.

3. The hydrogen ions may combine with ammonia. The distal tubular cells contain large amounts of an amino acid known as glutamine. When large amounts of hydrogen ion are being secreted, glutamine can split up to give glutamic acid and ammonia. The ammonia enters the tubular fluid where it combines with hydrogen ions to give ammonium ions (NH_4).

TUBULAR SECRETION

Most substances are removed from the body by the kidneys by glomerular filtration followed by a failure to reabsorb them in the tubules. Some substances are however actively secreted by the tubules from the blood into the urine. The most important mechanisms are:

1. The secretion of hydrogen ions into the proximal tubule.

2. The secretion of hydrogen and potassium ions into the distal tubule.

3. The secretion of a group of organic acids and other materials into the proximal tubule. This group includes penicillin, creatinine, phenol red, para-aminohippuric acid (PAH) and various iodine-containing substances used in renal radiology. Many of these substances are foreign to the body and it is difficult to see why there should be such a powerful mechanism for their secretion. It is probable that most of the foreign substances have some chemical grouping in common with a natural substance excreted by this route.

4. A group of strong organic bases, including guanidine and histamine are excreted by the proximal tubule.

HORMONAL CONTROL OF THE KIDNEY

Two hormones, vasopressin or antidiuretic hormone (ADH) and aldosterone are established as important controllers of renal function. Recent work suggests that other hormones such as growth hormone, prolactin and oxytocin may also be important.

ANTIDIURETIC HORMONE

This is secreted by the posterior lobe of the pituitary gland. The posterior pituitary is quite different from the anterior pituitary in that it has a very rich nerve supply. The nerve cells have their cell bodies in the part of the brain known as the hypothalamus and their axons travel down the pituitary stalk to the posterior pituitary. ADH is actually manufactured by the nerve cells in the hypothalamus and then passes down the axons to the posterior pituitary where it is released when nerve impulses pass along these fibres. The action of ADH is on the distal tubule and collecting ducts. In its absence, as far as water is concerned, these structures behave like steel tubes: they will not allow water to pass from the tubular fluid into the blood. The fluid which reaches the distal tubules from the loop of Henle is fractionally less concentrated than plasma as far as ions are concerned. In the

absence of ADH, when sodium and other substances are reabsorbed in the distal tubules water cannot follow and so must pass on to the urine. As the solids are removed the urine therefore becomes more and more dilute. There

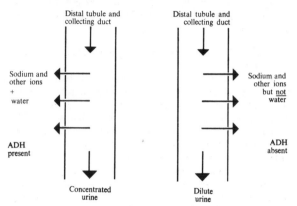

Figure 9.5 The action of antidiuretic hormone.

is a disease known as diabetes insipidus, the Latin name literally indicating the production of large amounts of tasteless urine. This disease occurs when little or no ADH is produced because of some malfunction of the pituitary or hypothalamus. As much as 10–20 litres of very dilute urine may be poured out every day. The patient must of course drink enormous quantities in order to maintain the water content of his body.

In contrast, when ADH is present in large quantities the distal tubules and collecting ducts are completely permeable to water. This means first that when any solid material is reabsorbed from the urine into the blood, an equivalent amount of water follows thus ensuring that the urine does not become more dilute than the plasma. Second, as the urine passes into the distal tubules and collecting ducts down through the medulla, it travels through regions where the interstitial fluid and cells have a steadily increasing concentration which in the inner parts of the medulla may be 3–4 times higher than the ionic concentration of the plasma. The concentration of the fluid outside the ducts is therefore greater than that inside and if the duct walls are permeable to water, water will move out of the ducts in order to maintain osmotic equilibrium. The net result is that in the presence of ADH the urine comes into equilibrium with the interstitial fluid through which it is passing and eventually becomes 3–4 times more concentrated than the plasma. By adjusting the concentration of ADH in the blood, the permeability of the distal tubules and collecting ducts can be adjusted to any desired level between complete permeability to water and complete impermeability. Thus urine of the desired concentration can be produced.

Several different factors govern the output of ADH:

1. The osmotic pressure of the blood passing through the hypothalamus where there are receptors for monitoring

it. If the blood becomes more dilute its osmotic pressure falls: this suggests that there is too much water in the body and the output of ADH is reduced. If the blood becomes more concentrated its osmotic pressure rises and ADH output is increased in order to keep water in the body.

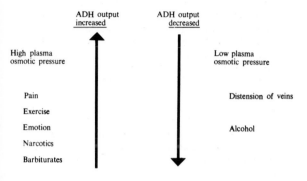

Figure 9.6 Factors influencing ADH output.

2. The degree of distension of the great veins. Stretch receptors in the thin-walled veins give a rough indication as to whether the circulatory system is overfilled with blood. If it is overfilled, ADH output is reduced in order to allow fluid to escape from the body.

3. Pain, exercise and emotion all increase ADH output, possibly because in these situations the rapid accumulation of fluid in the bladder is inconvenient.

4. Some drugs, notably alcohol, suppress ADH output and cause a diuresis (an increase in urine flow).

5. Other drugs, notably narcotics and barbiturates, increase ADH output and reduce urine volume.

ALDOSTERONE

Aldosterone is a steroid hormone manufactured by the outermost layer of the adrenal cortex. Its action is on the distal tubule where it increases the reabsorption of sodium and therefore, in the presence of ADH, of water as well. As the sodium is reabsorbed, potassium and

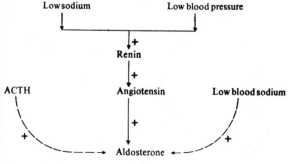

Figure 9.7 Possible factors influencing aldosterone output.

hydrogen ions tend to be excreted in exchange. The factors which govern the amount of aldosterone secreted are not yet fully clear and are the subject of much research. The present position is outlined in fig. 9.7.

PROLACTIN AND GROWTH HORMONE

It has recently been demonstrated that both prolactin and growth hormone can act on the kidneys to reduce the excretion of water, sodium and potassium. This action has been little investigated as yet but it may be important during the menstrual cycle and during pregnancy and lactation. Oxytocin may also have important renal actions.

INDICATIONS OF RENAL FUNCTION

It is helpful to have some relatively simple indicators of the extent of renal damage in patients with kidney disease. The simplest thing of all is to look at the urine and to check it for the presence of red cells and protein. Both are normally absent and either if present usually indicates some form of renal or urinary tract damage.

Another useful test is the estimation of the blood urea concentration. Urea is the end product of the breakdown of the nitrogen-containing parts of amino acids. It is formed from ammonia by the liver and excreted almost exclusively in the urine. It is freely filtered at the glomeruli and is not actively reabsorbed: some urea is passively reabsorbed along with the water, but large amounts are excreted. If the glomerular filtration rate falls, the rate of excretion of urea also falls and the blood concentration rises. However, there is a wide safety margin and the blood urea concentration does not rise sharply until the glomerular filtration rate has fallen to well under half its normal level. As the amount of renal damage increases beyond this point however, the blood urea level rises rapidly.

CLEARANCE

A somewhat more precise indication of the amount of renal damage that has occurred may be obtained by the

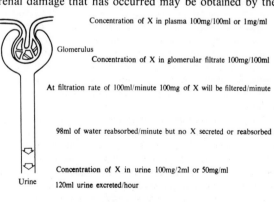

Concentration of X in plasma 100mg/100ml or 1mg/ml

Glomerulus

Concentration of X in glomerular filtrate 100mg/100ml

At filtration rate of 100ml/minute 100mg of X will be filtered/minute

98ml of water reabsorbed/minute but no X secreted or reabsorbed

Concentration of X in urine 100mg/2ml or 50mg/ml

Urine   120ml urine excreted/hour

$$\text{Clearance} = \frac{UV}{P} = \frac{50 + 120}{1} = 6000\text{ml/hr or }100\text{ml/min.}$$

where U   urine concentration of X in mg/ml
       V   volume of urine excreted in a given time
       P   plasma concentration of X in mg/ml

Figure 9.8 The concept of clearance.

use of the concept of clearance. We know that the concentration of a freely filtered substance is exactly the same in both the glomerular filtrate and the plasma. Suppose there were a substance which, once it had been filtered, was neither reabsorbed nor secreted during passage of the urine along the rest of the nephron. If this happened then all the substance filtered would appear in the urine. By collecting the urine over a period of time and analysing it, it would therefore be possible to know how much of the substance had been filtered by the glomeruli during that period. Since the concentration of the substance would be the same in both plasma and glomerular filtrate, it would be possible by taking a blood sample to know the concentration of the substance in the glomerular filtrate. Now the total amount of substance X filtered in a given time must be equal to the concentration of X in the glomerular filtrate multiplied by the volume of the filtrate in that time.

i.e.   Amount of X filtered = Concentration of filtrate × Volume of filtrate.

By analysing the urine and measuring its volume we can know the total amount filtered, by analysing the blood sample we know the concentration in the glomerular filtrate and it is therefore possible to calculate the volume of fluid filtered during a given time. The glomerular filtration rate gives a rough indication of how many glomeruli are functioning normally.

A number of substances are used for these 'clearance' studies as they are called. One which is natural in the body is creatinine, a breakdown product of the substance creatine which is found in muscles. The concentration of creatinine in the plasma is remarkably steady in any one individual and it is not reabsorbed at all by the nephron. It is secreted to a small extent and this means that more creatinine appears in the urine than was filtered at the glomeruli. In consequence, if creatinine is used for a clearance study the glomerular filtration rate is overestimated a little. However, because the creatinine clearance test is so easy to perform it is used frequently. A 12 or 24 hour collection of urine is made and the total amount of creatinine in it is estimated. During this period a blood sample is taken in order to estimate the concentration of creatinine in the plasma and in the glomerular filtrate and the filtration rate is estimated.

There are a number of other substances such as inulin (a carbohydrate) which are not normally found in the body but which when infused into the blood are freely filtered at the glomeruli and are neither reabsorbed nor secreted. These substances can therefore be used to estimate glomerular filtration rate relatively accurately. However they must be steadily infused during the test in order to maintain constant plasma concentrations and this makes them more difficult to use than creatinine.

The concept of clearance can also be used in the estimation of renal plasma flow. Suppose that there was a substance which was so effectively removed from the blood by a combination of glomerular filtration and tubular secretion that none was left in the renal vein i.e. all of the substance entering the kidney by the renal artery was removed from the blood during a single passage through the organ. If this were so, then the amount of the substance appearing in the urine in one minute (UV where U = the concentration of the substance in the urine and V = the urine minute volume) must equal the amount entering the kidney by the renal artery in one minute. If the plasma concentration of the substance was determined by withdrawing a venous blood sample and analysing it, then the renal plasma flow could be easily determined by dividing the total amount passing through the kidneys in one minute (UV) by the plasma concentration (P).

In practice no substance is totally removed from the blood during one passage through the kidney but several are more than 90% removed and can therefore be used to give a reasonable estimate of renal blood flow. Para-aminohippuric acid (PAH) is perhaps the best known of these.

MICTURITION

Urine is conveyed from the renal pelvis to the bladder by the ureter. The ureters are muscular tubes which contract rhythmically in waves so forcing the urine along. Where each ureter reaches the bladder there is a valve mechanism which allows urine to enter the bladder but not to pass in the reverse direction. This is important because it means that infection of the bladder (cystitis) cannot normally be transmitted backwards up to the kidneys. This is not true in pregnancy: for unknown reasons, but possibly because of the presence of high levels of progesterone, the smooth muscle of the ureters and the valves is in a very relaxed state. This means that especially during contraction of the bladder the urine can pass backwards up towards the kidneys.

The wall of the bladder is made up primarily of criss-crossing smooth muscle fibres. These fibres of the wall (the detrusor muscle) are relaxed except when micturition is actually taking place. Another smooth muscle, the internal sphincter, guards the exit from the bladder. It is

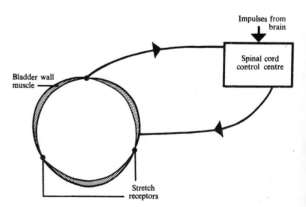

Figure 9.9 The control of micturition.

aided by a skeletal-type striated muscle, the external sphincter. Both sphincters are normally tightly contracted to prevent the escape of urine.

In the wall of the bladder there are stretch receptors which send information about the degree of fullness of the bladder to a control centre in the lowest part of the spinal cord. When filling reaches a critical level, the control centre, using both sympathetic and parasympathetic effector nerves, reverses the resting pattern of bladder muscle activity. The sphincters relax and the detrusor contracts, forcing out the urine. In paraplegic patients and in babies the reflex is not under conscious control and occurs automatically. In normal children and in adults the reflex can be blocked or activated by nerve impulses descending the spinal cord from the brain. These impulses may act on the control centre and may also alter the state of bladder receptors. If the smooth muscle adjacent to the receptors contracts, they fire off impulses just as they do when they are stretched by filling of the bladder. Relaxation of the smooth muscle reduces the impulse activity. Thus the brain can control micturition by modifying the activity both of the receptors in the bladder which supply the information and also of the control centre itself.

10 The Alimentary Tract

The main function of the alimentary tract is the transfer of food materials from the outside world to the blood. In the process, three separate stages are involved:

1. The food must be moved along the gut.

2. The food must be broken down into small molecules (digested).

3. The food must be transferred across the gut wall into the body (absorbed).

OUTLINE OF STRUCTURE

The gut throughout its length consists of inner mucosal and outer muscular layers. The function of the mucosal

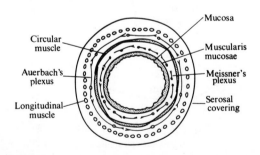

Figure 10.1 The layers of the gut wall.

layer is to produce secretions which lubricate and digest the food and to absorb the digested material. The function of the muscular layer is to propel the food along. The mucosa and the muscular layer are separated by an extremely thin layer of smooth muscle known as the muscularis mucosae. Outside this are much thicker inner circular and outer longitudinal layers of muscle. Complex nerve plexuses, usually known as Meissner's and Auerbach's, lie between the muscle layers.

NERVE SUPPLY

Three main types of nerve fibre supply the gut:

1. Sensory nerves which carry information about stretch of the gut wall and about the pH and chemical composition of the gut contents.

2. Parasympathetic nerves which release acetylcholine and which are responsible for stimulating the gut movements which propel the food along. The vagus supplies the gut up to the middle of the large bowel. The last part of the large bowel is supplied by parasympathetic nerves from the sacral region of the spinal cord.

3. Sympathetic nerves whose primary importance is in the control of the blood supply. They may also stimulate the action of the smooth muscle sphincters which guard the exits from and entrances to each section of gut.

The plexuses of nerves within the gut wall can do a remarkable amount by themselves and even when all the extermal nerves are cut can still to some extent coordinate the movement of food.

SMOOTH MUSCLE

The gut smooth muscle contracts spontaneously but the spontaneous activity can be modified by nerve activity and by circulating adrenaline. Acetyl choline released from parasympathetic nerves increases the activity of the main muscle in the gut wall and relaxes the sphincters. Adrenaline (epinephrine) relaxes the main gut muscle and stimulates the sphincters to contract, so stopping the movement of food. The part played by the sympathetic nerves is uncertain, but they may stimulate the sphincters to contract.

The movements of gut smooth muscle are of two main types:

1. Churning movements (segmentation) in which the muscle contracts and relaxes mixing the gut contents but not pushing the food along very much.

2. Propulsive movements (peristalsis), in which a coordinated wave of relaxation followed by contraction passes along the intestine. Peristalsis may be initiated by stretch of the gut wall by a ball of food. There is then reflex relaxation in front of the ball and reflex contraction behind it, so pushing the food along. Extrinsic nerves are not essential although they do facilitate the process.

THE MOUTH AND OESOPHAGUS

In the mouth the food is chewed to break it up into small pieces and to lubricate it by mixing it with saliva. The saliva also dissolves some of the food so allowing it to be tasted. Saliva contains an enzyme, amylase, which can begin the conversion of starch to the disaccharide, maltose. However this digestive function is probably not very significant: its main purpose is probably to clear the mouth of fragments of food. Cessation of salivary secretion leads to foul breath within an hour and dental decay within a week. Saliva is an alkaline, watery fluid, rich in sodium, potassium, chloride and bicarbonate. It also contains mucus which acts as a lubricant. The mucus and the watery secretion are produced by different cells. There are four main types of salivary gland:

1. *Tiny groups of cells* on the surfaces of the mouth and pharynx which secrete a mucus-rich fluid.

2. *The sublingual gland* beneath the tongue contains mainly mucus-secreting cells.

3. *The parotid gland* on the side of the face in front of the ear contains cells which secrete a watery, enzyme-rich fluid.

4. *The submandibular glands* tucked underneath the jaw contain both types of cells.

Secretion of mucus is brought about mainly by sympathetic nerves while parasympathetic impulses cause a profuse flow of watery saliva. Several different sorts of sensory stimulus may lead to activation of these nerves and salivary gland secretion. The main ones are:

1. Smelling or anticipating food as shown by Pavlov's dogs.

2. The act of chewing.

3. Chemical stimulation of taste receptors by food in the mouth. Bitter substances are by far the most effective.

SWALLOWING

Swallowing may be initiated either voluntarily or reflexly by touching certain areas near the back of the mouth. Anyone who has sat in a dentist's chair knows the power of the reflex and the difficulty of controlling it once initiated. The control centre for the reflex is in the lowest part of the brain, the medulla oblongata.

Swallowing begins with the tongue pressing up against the hard palate (the roof of the mouth) and pushing a ball of food back into the pharynx at the back of the mouth. Once the food is in the pharynx all voluntary control of the reflex is lost. Muscles in front of the food relax while those behind it contract. The wave of relaxation and contraction pushes the food down into the stomach. The movement of both food and fluids depends on muscular activity rather than on gravity and it is possible to hang upside down and to swallow effectively.

The entry to the stomach is guarded partly by muscle in the wall of the oesophagus and partly by the action of the diaphragm which nips the oesophagus. These mechanisms normally prevent the stomach contents moving backwards up the oesophagus. If because of some abnormality or some temporary malfunction the acid stomach contents do pass up into the oesophagus, the acid stimulates pain receptors giving the familiar sensation sometimes known as 'heartburn.'

THE STOMACH

An outline of structure is shown in fig. 10.2. The stomach

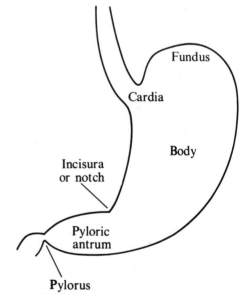

Figure 10.2 Outline of stomach structure.

is a large muscular bag which has the following functions:

1. To store food and pass it on to the intestines in a controlled, steady manner.

2. To begin the digestion of proteins.

3. To mix up and soften the food by its regular churning movements.

4. To secrete 'intrinsic factor' which is essential if vitamin B12 is to be absorbed. The intrinsic factor is a

mucoprotein which combines with the vitamin. The combination, but not the vitamin alone, can then be bound to the wall of the intestine and the vitamin taken into the body.

5. To absorb a few substances. In general the stomach is not a major organ of absorption but some substances such as alcohol and aspirin are relatively rapidly absorbed by it.

STOMACH MOVEMENTS

Spontaneous waves of activity begin near the top of the stomach where the oesophagus enters and move regularly down to the pylorus about three times per minute. Hormones and nerve activity do not alter the rhythm but they do alter the intensity of the contractions. Acetyl choline released by the vagus fibres to the stomach is by far the most important factor in maintaining normal gastric movements. Section of the vagus as part of the surgical treatment of peptic ulcers severely interferes with the normal emptying of the stomach: it must be combined with a so-called 'drainage' operation. In such an operation the exit of food from the stomach is made easier either by widening the pylorus (pyloroplasty) or by making an additional exit directly from the bottom of the stomach to the small intestine (gastro-enterestomy).

When the stomach has recently been emptied of food, the contractions are very feeble. As hunger develops, they increase in intensity but then become very weak again when the stomach is filled: gradually they become stronger and each one then forces a spurt of food into the duodenum.

Several mechanisms, all starting with the activation of sensory receptors in the duodenum, are important in modifying the rate of stomach emptying. Chyme (the mixture of food from the stomach) in the duodenum slows down the rate at which food enters the duodenum particularly if the chyme is acid or fatty. Three mechanisms are involved:

1. The receptors initiate impulses which cause the release of a hormone from the duodenal wall. This is known as enterogastrone and it circulates in the blood and reduces stomach activity.

2. The receptors send information to the central nervous system which then reduces the activity of the vagus nerve, again diminishing the intensity of stomach contractions.

3. Receptors send impulses to the stomach along local nerves.

GASTRIC SECRETION

The main part of the stomach surface is lined by simple cells which secrete mucus and a small amount of an alkaline fluid which helps to protect the stomach from the action of its own gastric juice. Leading down from the surface are many deep pits, the gastric glands, which contain the cells which secrete the gastric juice. There are three main types of cells:

1. Those near the neck of the gland secrete mucus which spreads over the surface of the stomach as a protective layer, trapping the alkaline fluid and preventing acid and enzymes getting at the stomach surface cells.

2. Deeper in the gland are chief or zymogen cells which produce an inactive enzyme precursor known as pepsinogen.

3. Also deep in the gland are the parietal or oxyntic cells which secrete hydrochloric acid and possibly intrinsic factor.

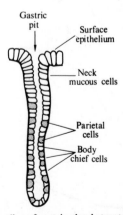

Figure 10.4 An outline of gastric gland structure.

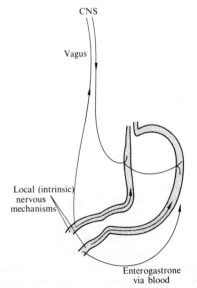

Figure 10.3 The control of stomach movements by the entry of food into the duodenum.

The glands in the cardiac area where the oesophagus enters the stomach contain mucus-secreting cells only.

Those in the pyloric area near the exit contain mucus-secreting and enzyme-secreting but no acid-secreting cells. Those in the main body of the stomach contain all three cell types.

Pepsinogen is secreted in an inactive form because if it were made in active form it would break up the proteins in the secreting cells themselves. It is activated by conversion to pepsin on contact with acid: the acid splits off a small part of the pepsinogen molecule so exposing the active part of the enzyme. Once some pepsin has been formed, it itself can then convert pepsinogen to pepsin. The pH of the normal stomach contents is 2–3 because this is the optimum pH for the protein-splitting activity of pepsin. Little if any digestion of carbohydrates or fats takes place in the stomach.

Stimuli to secretion

The secretion of gastric juice is conventionally divided into three phases which merge into one another.

1. *Cephalic phase*. This refers to the gastric secretion which can be brought about when no food has yet entered the stomach. On smelling, seeing or thinking of food, and especially on putting food into the mouth, there is an outpouring of gastric juice. The information from the receptors is carried to the control centre in the medulla oblongata in the lower part of the brain. This activates the vagus which releases acetyl choline and which stimulates the gastric glands to secrete. This phase is known as the cephalic phase because it depends on the control centre in the brain.

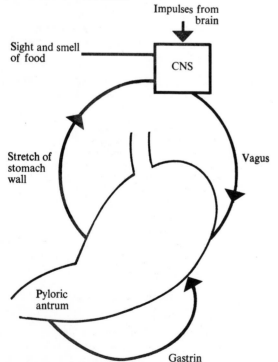

Figure 10.5 The control of gastric secretion.

2. *Gastric phase*. This refers to the secretion which occurs in response to the actual entry of food into the stomach. Chemical and stretch receptors in the stomach wall, particularly in the pyloric region, lead by three mechanisms to an increase in the flow of gastric juice.

a. By the intervention of the nerve plexuses in the stomach wall.

b. By sending impulse to the control centre in the CNS which then increases the activity of the vagus.

c. By causing the release of a hormone called gastrin from the region of the stomach known as the pyloric antrum. Gastrin enters the venous blood, is carried to the heart and pumped out by the arteries again to the stomach. On reaching the gastric glands it powerfully stimulates their secretion.

3. *Intestinal phase*. If in an experiment food is placed directly into the small intestine without first passing through the stomach, gastric secretion is stimulated. Thus the digestion of food in the stomach is promoted by the presence of food in the intestines. The mechanisms are not known.

IRON AND CALCIUM ABSORPTION

Although iron and calcium absorption do not take place in the stomach, the normal secretion of acid gastric juice seems to be essential for the absorption of these ions in the intestine. Individuals whose acid secretion is reduced either because of gastric atrophy or because of surgical or medical attempts to treat peptic ulcers, may fail to absorb calcium and iron normally and eventually become deficient in these ions. The precise mechanism of this effect is uncertain but it seems probable that the acidity helps to maintain phosphate in the more soluble $HPO_4^=$ form and also in the conversion of ferric iron to ferrous iron, the form in which it must be absorbed.

THE PANCREAS

The pancreas is the single most important source of digestive juices. Its secretion contains enzymes which can split up all three of the major foodstuffs, fat, carbohydrate and protein. The enzymes all work best at a pH which is slightly on the alkaline side of neutral. They will not work at the acid pH of gastric juice and so it is important that the gastric secretions should be neutralised. This is achieved partly by the bile but mainly by the large amounts of bicarbonate present in the pancreatic secretions.

The pancreas is, of course, also an important endocrine gland because it contains the islets of Langerhans which secrete insulin and glucagon. This aspect of its functon is discussed in chapter 5.

The cells which secrete the digestive juices are arranged in groups called acini which empty into tiny tubules. These join up until eventually they form the main pancreatic duct which empties into the duodenum. It often shares a common emptying point with the bile duct. There is a powerful sphincter which controls the emptying of bile and pancreatic juice into the duodenum.

The vagus nerve and the hormone gastrin can both stimulate pancreatic secretion but they are probably much less important than two other hormones, secretin and pancreozymin. Secretin causes a copious flow of watery juice while pancreozymin stimulates the secretion of an enzyme-rich juice. Normally the two act together. They are both manufactured in the wall of the duodenum and are both released into the blood when food enters the duodenum from the stomach. The hormones are then carried by the blood around to the pancreas. Acid is a potent stimulator of the flow of secretin while partially digested foods, especially proteins and fats, stimulate the flow of pancreozymin.

THE BILE

Bile has two main functions. It acts as an excretory channel for certain substances, notably cholesterol and the breakdown product of haemoglobin, bilirubin, and it also contains the bile acids which are important in the digestion of fat. These functions are discussed in chapter 3. This section is concerned with the control of bile secretion.

Bile is manufactured in the liver and between meals it is stored in the gall bladder. In the gall bladder, sodium, chloride, bicarbonate and water are all absorbed from the bile into the blood. Cholesterol, bilirubin and the bile salts are all left behind and because of the absorption of water they may become 5–10 times more concentrated than in the bile freshly secreted by the liver. Partly as a result of this concentration, some of the solid material (especially cholesterol and to a lesser extent bilirubin and other substances formed from it which are collectively known as the bile pigments) tends to precipitate out giving gall stones which may sometimes escape from the gall bladder and block the bile ducts. However concentration cannot be the only factor in the formation of gall stones since it occurs in everyone while gall stones, although common, are by no means universal.

Control of the flow of bile into the duodenum depends on two factors, control of the rate of bile secretion by the liver and control of the emptying of the gall bladder. The main factors which alter the rate of bile secretion are probably the following:

1. Activity of the vagus nerve.

2. The hormone secretin, released by the duodenum when food enters the duodenum from the stomach.

3. The presence of bile salts in the blood carried to the liver by the portal vein.

About half an hour after feeding, the gall bladder contracts and pours its bile into the duodenum just as food is beginning to enter the small intestine from the stomach. This contraction depends on two mechanisms:

1. Food in the duodenum stimulates sensory receptors which then lead to reflex activity in the vagus. The vagal

fibres to the gall bladder release acetyl choline which causes muscle contraction.

2. Food in the duodenum causes release of a hormone called cholecystokinin from the duodenal wall. Cholecystokinin circulates in the blood and causes strong contractions of the gall bladder.

THE SMALL INTESTINE

The small intestine consists of the short duodenum leading on to the much longer jejunum and ileum. It secretes a digestive juice, sometimes known as the succus entericus, whose composition has not yet been precisely defined. The major functions of the small intestine are to provide a site where digestion can occur and to absorb the digested food. The biochemistry of digestion and absorption is dealt with in chapter 3.

In order to carry out its functions, the small intestine must be able to churn the food up and move it along and must be able to provide good conditions for absorption. The movements are similar to those which occur in the rest of the gut and they have been described at the beginning of the chapter.

ABSORPTION

Absorption of the digested food into the body must take place through the gut wall. Since the greater the surface area of the gut wall the more rapidly absorption can take place, it is not surprising that the surface area of gut available is increased in three different ways:

1. The whole mucosal surface is thrown into folds.

2. The mucosal folds are covered by minute finger-like projections known as villi: each villus has a rich blood supply and also a large central lymphatic known as a lacteal. Most of the digested food passes into the blood, but the fats tend to enter the lacteals.

3. Each cell surface facing the gut is itself covered with myriads of tiny projections known as microvilli. These greatly increase the surface area available.

In the cells which line the gut there are active mechanisms for moving many substances into the body. Those which are actively shifted in this way include sodium, glucose and amino acids. The movement of these solids tends to leave the gut contents of a lower osmotic pressure than the plasma and so water automatically follows them into the blood. Some other absorptive mechanisms which are not infrequently affected in disease are:

1. *Calcium.* The absoption of calcium depends on gastric acidity and on the availabilty of adequate supplies of vitamin D. The way in which the acid operates is unknown but it may help to shift the calcium into a more soluble form. Patients who have defective acid secretion, either naturally or after surgery for peptic ulcer, tend to become calcium deficient. If there is too much phosphate

in the food the calcium may be precipitated in a very insoluble form which cannot be absorbed. This sometimes occurs in new born infants fed cows' milk which has a much higher phosphate content than human milk.

2. *Iron*. The normal absorption of iron, again for uncertain reasons, requires the secretion of normal amounts of gastric acid and absorption is reduced if secretion is defective.

3. *Fat soluble vitamins* (A, D, E and K). The absorption of these depends on the normal absorption of fat. If anything interferes with fat absorption (e.g. inadequate bile or pancreatic juice secretion), then vitamin deficiencies may follow.

4. *Vitamin B_{12}*. As mentioned at the beginning of the chapter, the absorption of vitamin B_{12} depends on the secretion of a mucoprotein 'intrinsic factor' by the stomach. The intrinsic factor combines with the vitamin. The complex is then bound by the wall of the ileum and the vitamin is absorbed. In the absence of intrinsic factor the disease known as pernicious anaemia develops.

Absorption of all valuable solid materials is complete by the time the end of the ileum is reached. There is a valve which guards the exit from the ileum to the large intestine. This is important because the contents of the small intestine are normally sterile and creamy-white in colour. Bacterial infection of the small intestine due to reflux from the colon may lead to serious problems of absorption. The small intestine normally delivers 400–600 ml of material to the large intestine every day. Most of this consists of water, indigestible material and remnants of dead cells which are continually being shed from the gut surface.

LARGE INTESTINE
The large intestine has three main functions:

1. It offers a place for the storage of unwanted material until it can be excreted as faeces at a convenient time.

2. It absorbs water so that the fluid material which leaves the small intestine (chyle) is reduced to about 25% of its volume by the time it is excreted as faeces.

3. It offers a home to a large colony of bacteria. It is being increasingly recognised that these bacteria synthesise vitamins and may be important sources of vitamins of the B group, especially folic acid.

The large intestine consists of a blind sac, the caecum with its attached appendix, the long ascending, transverse and descending colons and the rectum where the faeces is finally stored.

DEFAECATION
Defaecation is stimulated by the stretching of receptors in the wall of the rectum. This stretching may arise from two sources:

1. Distension of the rectum with faeces.

2. Contraction of the muscle of the rectal wall. The receptors behave as though they are fixed between the ends of the muscle fibres and so when the muscle fibres contract the receptors between them are stretched. The rectal wall muscle may be stimulated by voluntary action or, as in diarrhoea, by the presence of abnormal bacteria or toxins. The control mechanism is described fully in chapter 2.

In infants or in older people who have lost control of defaecation because of damage to the spinal cord, distension of the rectum to a certain critical level automatically fires off the defaecation reflex. There is a control centre in the lower part of the spinal cord which orders the following effector acts when the sensory input to it reaches trigger point:

1. Peristalsis occurs in the large intestine forcing faeces along to the rectum.

2. The muscles in the rectum contract, the circular ones raising pressure and the longitudinal ones pulling the rectum up and over the faeces it contains.

3. The internal (smooth muscle) and external (striated muscle) sphincters relax and the faeces are automatically expelled.

In normal individuals the reflex may be voluntarily suppressed by impulses coming down the spinal cord from the brain to the control centre and also by relaxation of the rectal muscle which takes the tension off the receptors. On the other hand, when the time is appropriate, the reflex may be voluntarily encouraged by instructions sent to the control centre and by contraction of the rectal muscle which stimulates the receptors. The expulsion of faeces may also be voluntarily assisted by raising the intra-abdominal pressure by means of the Valsalva manoeuvre. This familiar act consists of closing the glottis, pulling the diaphragm down, and contracting the muscles of the abdominal wall. The complete voluntary control of defaecation depends on an intact spinal cord and on intact somatic and parasympathetic peripheral nerves.

A great deal of nonsense is talked about the need to empty the bowels once per day. It is true that this is the most usual pattern, but there is immense individual variation: some entirely normal individuals may empty their large intestines once a week and others three times a day. Neither pattern need be abnormal provided that it is the usual one for that individual. A *change* in bowel habit, however, is a very important clinical feature as not infrequently it indicates disease.

The discomfort which is associated with constipation is primarily due to stretching of the large intestine and has little if anything to do with the absorption of toxins. The sensation may be rapidly produced by placing a balloon in the rectum and blowing it up and may be instantly relieved by letting down the balloon.

CLINICAL ASPECTS OF GUT FUNCTION

Most disorders of gut function can be more easily understood if the physiology is known. Three of the more important aspects of disordered physiology will be discussed here.

FLUID LOSSES

Fluid losses from the gut are not uncommon. Vomiting leads to a loss of gastric secretion while diarrhoea may lead to loss of the contents of the large intestine. Patients with an ileostomy may lose fluids from the small intestine (An ileostomy is a way of by-passing the large intestine if the latter is severely diseased or has to be surgically removed: the end of the ileum is brought out through the anterior abdominal wall and empties into a bag). Excess fluid loss from any part of the gut leads to the loss of water, of sodium and accompanying negative ions, and of potassium. In some ways the potassium loss is the most important, partly because it tends to be forgotten, partly because the amount of potassium in the extracellular fluid is relatively small and partly because even small variations in plasma potassium concentration may have serious effects on heart function. It is very important to ensure that any fluid and ions which are lost are promptly replaced.

The effect of the fluid loss on acid-base balance depends on which fluid is lost. Usually fluid lost from the stomach is acid and so an excess of alkali is left behind (metabolic alkalosis). Fluid lost from below the duodenal level is usually alkaline and so leaves an excess of acid behind in the body (metabolic acidosis).

PEPTIC ULCERS

An ulcer occurs when epithelial tissue is lost from any surface in the body leaving a raw area. A peptic ulcer is an ulcer which occurs in a part of the gut exposed to 'peptic' (gastric acid-containing) juice: in effect this usually means ulcers in the stomach and duodenum.

An ulcer could occur either because of a lack of the normal factors which protect the gut surface or because of an excess of the powerful destructive components of gastric juice. Mucus is probably the most important protective agent while acid and pepsin are the destructive ones. There is a lack of agreement among research workers as to what is most important but tests of acid secretion have yielded interesting results. Acid secretion may be tested by putting a tube into the stomach of a resting person who has not eaten for 12 hours or more. The contents are sucked out and then suction is con-

tinued for an hour in order to estimate the so-called basal rate of acid secretion. An injection of histamine (preceded by an anti-histamine to avoid unpleasant side effects) or of a synthetic relative of gastrin known as 'pentagastrin' is then given. The acid is then collected for another hour giving an estimate of the maximal rate of secretion. These tests have now been carried out in very large numbers of patients and it has been found that duodenal ulcer patients tend to have high acid secretion rates while gastric ulcer patients tend to have low acid secretion rates. This suggests that duodenal ulcers may be due to an excessive secretion of gastric juice while gastric ulcers may be due to defective defence mechanisms. The idea that gastric ulcers are associated with defects in the stomach wall is supported by the fact that gastric ulcers are associated with gastric cancer and with pernicious anaemia due to lack of secretion of intrinsic factor. Duodenal ulcers are not associated with either cancer or pernicious anaemia.

MALABSORPTION

Malabsorption, as the word implies, is a condition in which the food passing through the gut is not absorbed normally. Large amounts pass straight through the small intestine into the large bowel and out in the faeces, causing diarrhoea and malnutrition. The condition has two fundamental causes:

1. *A failure of digestion.* If food is not digested it cannot be absorbed even though the absorptive mechanisms of the small intestine are working normally. Since the most important of all the digestive juices comes from the pancreas, pancreatic disease frequently causes malabsorption. Lack of normal bile secretion causes a failure of fat absorption. There are also a number of diseases in which enzymes important in digestion are congenitally absent. These are most important in infancy and are discussed in chapter 3.

2. *A failure in the absorptive process itself* because of damage to the wall of the small intestine. The most important causes of this are coeliac disease and sprue. Coeliac disease is in part a failure of digestion because the gut seems incapable of digesting fully the protein, gluten, which is found in wheat. Some of the products of partial digestion are toxic and damage the wall of the small intestine causing atrophy of the villi and a failure of absorption. A complete cure can be achieved by withdrawing wheat products from the diet. In sprue the wall of the small intestine degenerates in much the same way as in coeliac disease but the cause seems to be quite different although it is by no means fully understood. One possibility is that a change in the types of bacteria present in the gut means that they consume folic acid instead of producing it. Sprue can usually be relieved by giving large doses of folic acid and by altering the gut bacteria again by the use of antibiotics.

11 The Reproductive System

Human reproduction, as everyone knows, is of the sexual variety. The process begins with the formation of eggs (ova) in the female ovaries and of spermatozoa (sperm) in the male testes. In the act of sexual intercourse, if it occurs at the right time, the two are brought together and the fertilised egg or zygote results. This becomes implanted in the uterus which has been prepared for its reception and develops over approximately 280 days in normal circumstances to a fully formed infant. In the process of labour (parturition) the infant is pushed into the outside world and the cycle normally ends with the feeding of the infant with milk from the mother's breasts.

MALE REPRODUCTIVE FUNCTION

The primary reproductive organs in the male are the testes. These are responsible for the production of both sperm and of male sex hormones (androgens), of which testosterone seems to be the most important. For normal function, the testis is dependent upon the hypothalamus and pituitary. The rest of the male internal and external genitalia, shown in outline in fig. 11.1, essentially consist

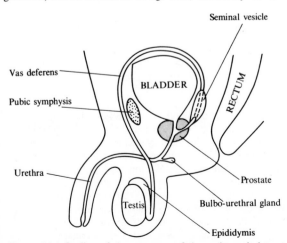

Figure 11.1 Outline of the structure of the male genital tract.

of mechanisms for delivering the sperm to the female in a healthy condition.

CONTROL BY PITUITARY AND HYPOTHALAMUS

Before puberty the testes are small and produce only minute amounts of testosterone. The testosterone circulates in the blood and its presence is detected by the part of the brain known as the hypothalamus. The hypothalamus is connected by means of the special system of blood vessels known as the pituitary portal system to the anterior pituitary. The anterior pituitary produces two hormones, the so-called gonadotrophic hormones, which can affect the testes. These are follicle-stimulating hormone (FSH) and luteinising hormone (LH): they were originally named because of their effects in the female but since it is now known that the male and female pituitary hormones are identical it has become conventional to call the two hormones FSH and LH whether one is discussing males or females. LH in the male is still sometimes known as interstitial cell stimulating hormone or ICSH.

Before puberty FSH and LH are secreted by the anterior pituitary in negligible quantities. This is because they depend for their secretion on two substances released from the hypothalamus into the pituitary portal system. These two are FSH-releasing factor and LH-releasing factor. Before puberty the releasing factors are secreted by the hypothalamus in only very small amounts: the minute quantities of testosterone produced by the testes seem able to prevent the hypothalamus from secreting larger amounts of the releasing factors.

PUBERTY AND THE FUNCTIONS OF THE TESTES

At puberty something which is not yet understood happens to the hypothalamus. It becomes much less sensitive to the testosterone in the blood and the minute quantities of androgen can no longer prevent the hypothalamus from secreting large amounts of the releasing factors. And so, in turn, the pituitary begins to secrete

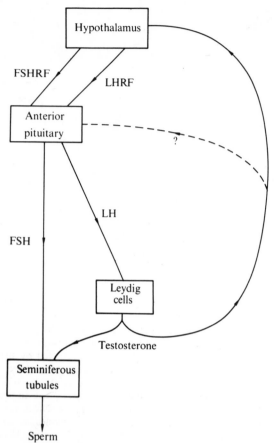

Figure 11.2 The relationship between the seminiferous tubules of the testis which produce sperm, the Leydig cells which produce testosterone, the hypothalamus and the anterior pituitary.

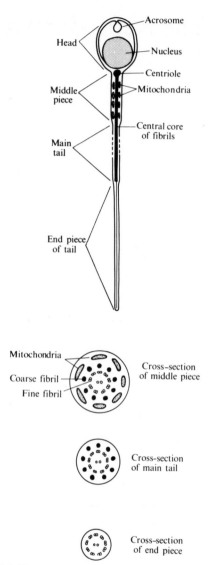

Figure 11.3 The structure of a sperm.

large amounts of FSH and LH. These two hormones circulate in the blood and act on the testes to convert them to the adult form.

The testes contain three main types of cell:

1. *Germ cells* involved in the production of vast numbers of sperm. During the cell divisions which occur, the number of chromosomes is reduced from the 46 found in all other cells in the body to the 23 found in a sperm. A similar process occurs during egg formation so that when a sperm containing 23 chromosomes fuses with an ovum containing 23 chromosomes, the result is a zygote with the normal number of 46 chromosomes. A mature sperm is a remarkable cell consisting primarily of a huge head which carries the nucleus with its chromosomes, a middle piece containing many mitochondria whose function is to supply energy and a long tail which utilises this energy to lash the fluid in which the sperm is bathed, so enabling it to swim.

2. *Sertoli cells* of uncertain function but which may supply nutrients for the sperm.

3. *Interstitial cells* (Leydig cells) which manufacture testosterone.

Broadly speaking, LH stimulates the secretion of testosterone and the testosterone acts with FSH to promote the development of mature sperm. FSH and LH together therefore stimulate the testis to grow and to change its function to the adult type. The hypothalamus still remains dependent on testosterone for controlling the amounts of releasing factors which it secretes, but very much larger amounts of testosterone are now required. However, if testosterone concentration in the blood becomes too high, FSH and LH secretion rates are reduced until testosterone concentration falls back to the desired level. If it becomes low, FSH and LH output are increased in order to return testosterone levels back to normal.

SECONDARY SEXUAL CHARACTERS

At puberty the increased output of testosterone causes the familiar changes which convert a boy into a man. The main ones are:

1. The internal and external genitalia and in particular the penis grow to adult size.

2. Because of changes in the vocal cords, the voice breaks and becomes deep.

3. The skin becomes coarser and much more grease is secreted.

4. The muscles develop and strengthen and the personality tend to become more aggressive.

5. Pubic and axillary hair grow. The male pattern of pubic hair has a point going up towards the umbilicus while the female pattern is a flat-topped triangle.

6. The hair line over the temples recedes to give the adult male pattern.

7. The fusion of the growing points of the bones (the epiphyses) is accelerated, so tending to terminate growth.

THE FEMALE SEXUAL CYCLE

Female sexual function is much more complex and much less well understood than sexual function in the male. The primary organs are the ovaries which have two main functions, the production of ova and the manufacture of hormones.

THE FORMATION OF OVA

At birth the ovary contains between 200,000 and 500,000 structures known as primordial follicles, each of which can give rise to a single mature egg. More than half of these degenerate during childhood but the 100,000 plus which are left at puberty are more than enough to supply one egg per lunar month for the 30–40 years of female reproductive life.

Before puberty no mature ova are released because the amounts of FSH and LH released by the pituitary are too small: the secretion of the hormones seems to be suppressed by tiny amounts of oestrogen (female hormone) from the ovaries acting on the hypothalamus. At puberty, as in the male, the output of FSH and LH rises, the ovary develops and fully formed eggs begin to be released at monthly intervals.

In an adult, at about the time of menstrual bleeding, several primordial follicles enlarge and the cells surrounding them (the membrana granulosa cells) multiply. Within a few days, for unknown reasons and by

unknown mechanisms just one of these primordial follicles is selected: it enlarges rapidly while the others which had been enlarging degenerate. A fluid filled cavity

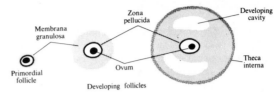

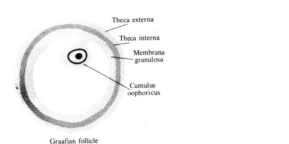

Figure 11.4 Outline of follicular development in the ovary (not to scale).

soon appears within the granulosa cells giving a structure known as a Graafian follicle. The cells surrounding the follicle, the theca interna and theca externa, secrete female hormones. The Graafian follicle grows and moves towards the surface of the ovary. Eventually, usually about 12–14 days after the beginning of the previous menstrual period, the follicle ruptures, releasing the ovum into the abdominal cavity, normally close to the opening of the Fallopian tube. The egg is then carried down to the uterus by the movements of the tube and of the cilia which line its walls.

The follicular cells which remain in the ovary after the egg has been released are rapidly vascularised by the growth of new blood vessels. Under the stimulus of the increased blood supply the cells grow rapidly. Many of them become yellow in colour and because of this the ruptured follicle becomes known as the corpus luteum. The corpus luteum is an active secretor of female hormones and remains functional until about the time of the next menstrual flow when it degenerates.

THE FEMALE HORMONES

As in the male, the primary sex organ, the ovary, secretes only very small amounts of hormone before puberty. After puberty, mainly under the influence of LH, the ovaries secrete much larger amounts of hormones which fall into two groups known as the oestrogens and the progestins. Oestradiol is probably the most important natural oestrogen and progesterone the most important natural progestin but there are many other substances in each group, both natural and synthetic.

The oestrogens are primarily responsible for the changes which occur at puberty. The main changes are:

1. Development of internal and external genitalia.

2. Growth of the breasts and rounding of the body contours due to selective deposition of fat.

3. Changes in psychological make up.

4. The beginning of menstruation (menarche).

5. Growth of pubic and axillary hair. Adrenal androgens may be more important than ovarian hormones in this, since hair growth is normal at puberty even when the ovaries are congenitally absent (Turner's syndrome).

Progesterone is the hormone which is secreted by the corpus luteum and its actions are discussed in the next sections.

The Hypothalamus, Pituitary and Ovaries

The control of hormone secretion is much more complex in the female than in the male and is relatively poorly understood. During the first half of the menstrual cycle until just before the egg is released (ovulation), the rate of oestrogen secretion rises steadily but little if any progesterone is produced. A few hours before ovulation actually occurs there is a sharp rise in the output of progesterone. During the second half of the cycle (luteal phase) both oestrogens and progesterone are present in relatively high concentrations but their levels fall again as the corpus luteum degenerates and menstrual flow begins.

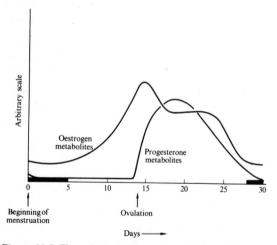

Figure 11.5 The urinary outputs of metabolites of oestrogen and progesterone during the menstrual cycle.

No one doubts that this pattern is dependent on pituitary hormones but most of the details remain obscure. The one outstanding feature is that there is a very sharp peak in the output of LH just around the time of ovulation. This peak in LH secretion almost certainly stimulates the release of progesterone and of the egg. In most women the peak and the subsequent ovulation occur at about the mid-point (12th–16th days after menstrual flow starts) of a normal four-week cycle. In women with irregular cycles

ovulation usually occurs about 14 days before the following menstrual flow. It seems to be the first part of the cycle which is irregular while the second almost invariably lasts for two weeks.

Some animals, notably the rabbit, do not have a cycle as such. The ovary is ready to release eggs at almost any time. When copulation takes place, the nervous stimuli associated with it cause a burst of LH release from the pituitary which then leads to ovulation. There is some evidence that a similar process can occur in some women for there are a number of reasonably well-documented cases where a single act of intercourse occurring just after or just before menstruation has been followed by pregnancy.

MENSTRUAL CYCLE

The ovarian cycle of changing rates of hormone secretion is followed by cyclic changes (produced by the hormones) in other organs. The most obvious changes occur in the uterus, resulting in the menstrual cycle. Most of the features of the menstrual cycle can be imitated by treatment with oestrogen alone for about two weeks, followed by oestrogen+progesterone for about 8–10 days and then by 5 days without hormones before the next oestrogen treatment begins.

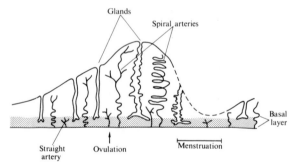

Figure 11.6 Changes in the endometrium which occur during the menstrual cycle.

The wall of the uterus is made up mainly of smooth muscle fibres and the cavity is lined by a special layer of tissue known as the endometrium. The endometrium is made of connective tissue. Its surface is covered by a ciliated epithelium and from the surface mucus-secreting glands dip down into the apparently structureless connective tissue which is known as the stroma. There are two sorts of arteries: simple straight ones supply the deepest layers of the endometrium which are not shed during menstruation, while tortuous spiral vessels supply the more superficial layers.

At the end of menstruation, all the superficial layers are shed. The first stage in rebuilding is a rapid re-covering of the raw surface with ciliated epithelium. The glands at this stage are simple and straight. During the first half of the cycle (the follicular phase) the endometrium thickens slowly, the spiral arteries develop and the deepest parts of the glands become dilated. These effects are due to oestrogen and can be imitated by giving oestrogen to women without ovaries; on stopping the oestrogen treatment the endometrium breaks down and bleeding occurs but it is much less profuse than in normal menstrual cycle.

During the luteal phase the thickening of the endometrium becomes much more rapid. The spiral arteries become tortuous and swollen with blood and the glands elongate rapidly becoming twisted and folded. These changes can be imitated in women without ovaries by following a course of oestrogen alone with one of oestrogen+progestin. On cessation of the hormone treatment, bleeding occurs but this time is indistinguishable from that of a normal period. These observations suggest that menstrual bleeding is naturally due to a fall in the blood levels of oestrogen and progesterone at the end of the menstrual cycle.

Other organs apart from the uterus are also affected by the ovarian hormones. The cervical canal which leads from the uterus to the vagina has walls which secrete mucus. Increasing oestrogen concentrations make the mucus thinner, more alkaline and more easily penetrated by sperm. It is thinnest at the time of ovulation. Rising progesterone levels make the mucus thick, acid and difficult to penetrate. Cyclic changes also occur in the vaginal secretions and mucosa but these are much less marked in humans than in many animals.

Just before menstruation many women become tense and irritable, concentrate less effectively and work less efficiently. The breasts enlarge and the tissues become loaded with salt and water. The reason for this 'premenstrual tension' is unknown although it is almost certainly caused by fluctuating hormone levels of some sort.

SEXUAL INTERCOURSE

Sexual intercourse is a complex series of events which requires the interaction of two individuals. Variations in pattern are almost infinite and only the most basic physiological details will be described here.

The first problem which must be overcome is the deposition of sperm within the female genital tract. The penis in its normal flacid state is a useless instrument for this purpose and so it must be converted into a stiff rod. This can be achieved because of the presence within the penis of three columns of spongy tissue, the corpora cavernosum. These are normally soft and relatively empty of blood. However, sexual excitement and particularly mechanical stimulation of the under side of the tip of the penis (the glans, which is Latin for acorn) sets off the reflex. Parasympathetic nerves to the penile arteries release acetyl choline which makes them open wide. Sympathetic nerves to the penile arterioles are inhibited and this allows the arterioles also to dilate. Blood at high pressure can therefore be pumped into the corpora cavernosa. The venous outflow is simultaneously compressed and so the penis becomes hard and stiff as it fills with blood.

During normal sexual play before intercourse, corresponding changes take place in the female, again brought about by reflexes initiated by sexual excitement and by mechanical stimulation of the breasts and genitalia. The most important feature in the female is a rapid increase in the flow from the vagina of a lubricating mucus-rich solution: this allows the penis to slip easily into the vagina. The clitoris and labia also becomes congested with blood and the nipples stand erect.

When both partners are ready, the erect penis is slipped into the lubricated vagina. During movements of the penis in and out, the frictional stimuli in both partners should ideally result in simultaneous orgasm. In practice, male orgasm virtually always occurs while a failure to reach female orgasm is quite common.

In the male, orgasm results in the ejaculation of sperm from the tip of the penis. This occurs because the smooth muscle in the epididymis, the vas deferens, the prostate gland and the seminal vesicles all contracts. Sperm are stored in the epididymis and vas and not in the seminal vesicles. The function of the vesicles is to secrete a fluid rich in fructose which bathes the sperm and supplies them with energy. The prostate gland secretes a complex fluid whose function is uncertain. The mixture of sperm, seminal vesicles fluid and prostatic fluid is known as semen. The normal ejaculate is 2–5 ml in volume and contains about 100 million sperm/ml. Frequent ejaculation reduces both the volume and the sperm count. Fertility seems to fall off when the count goes below 60 million/ml and successful fertilisation is unlikely if it is below 20 million/ml.

During orgasm and simultaneously with the contraction of the smooth muscle, the skeletal muscles of the pelvis and penis contract, so helping the expulsion of sperm. During intercourse the blood pressure and heart rate rise considerably and this may put some strain on the cerebral blood vessels, especially in older people. Some strokes caused by cerebral haemorrhage occur during sexual intercourse.

In females, orgasm is associated with uterine movements, probably stimulated by oxytocin released from the posterior pituitary gland. This is strongly suggested by the fact that orgasm in lactating women is commonly associated with ejection of milk from the breasts. The importance of these uterine movements is uncertain. They have been thought to be important in the transport of sperm but many women who never achieve orgasm obviously have no trouble in conceiving.

FERTILISATION AND SPERM TRANSPORT

It is believed that fertilisation normally takes place near the ovarian end of the Fallopian tube. The sperm therefore have a long way to travel from the vagina. In the vagina many may be killed because the vaginal secretions are often acid and this kills the sperm. The prostatic and seminal vesicle secretions are alkaline and tend to reduce the acidity. The cervix probably offers the most difficult barrier for the sperm to surmount and it is important that around the time of ovulation the cervical mucus should become thin. Once in the uterus, movements of the uterus and the Fallopian tubes, probably helped by cilia on the tube walls, assist the swimming sperm to reach their destination.

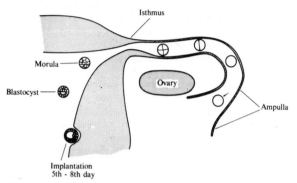

Figure 11.7 The female genital tract showing the probable site of fertilisation and the stages which the embryo goes through before implantation.

Fertilisation occurs when a single sperm penetrates and fuses with the ovum to give the zygote. The zygote soon begins to divide, becoming first a solid mass of cells (morula) in which a cavity later appears (blastocyst). The journey down the tube to the uterus probably takes 2–5 days after which for several days the embryo remains free in the uterine cavity. The outer layer of the embryo is known as the trophoblast and it is the trophoblast cells which burrow into the prepared endometrium and which stimulate the endometrium to form the maternal part of the placenta. This process is known as implantation.

PREGNANCY

The trophoblast of the implanted embryo differentiates into an outer layer where cell walls cannot be seen (syncytiotrophoblast) and an inner cellular layer (cytotrophoblast). Finger-like processes grow out from the trophoblast with blood vessels in their centres. These fetal blood vessels are connected to the main part of the fetus by the umbilical arteries and vein in the umbilical cord. On the maternal side of the placenta a cavity filled with blood develops into which the fetal villi project. Oxygen and food materials cross into the fetus from this lake while carbon dioxide and waste materials pass in the opposite direction.

HORMONES DURING PREGNANCY

If fertilisation occurs, the corpus luteum does not decay as in a usual cycle. Instead it continues to produce *oestrogen* and *progesterone* in order to maintain the uterine endometrium. How the ovary 'knows' that fertilisation has taken place is one of the unsolved mysteries of the reproductive process. It may have something to do with a hormone called *human chorionic gonadotrophin* (HCG) which during the first three months is secreted by the placental trophoblast in enormous quantities. It is probably HCG which stimulates the corpus luteum to

continue functioning. However, after 12–14 weeks of pregnancy the placenta itself secretes sufficient oestrogen and progesterone to maintain the pregnancy. The corpus luteum degenerates and, if necessary, the ovaries can be removed without terminating the pregnancy. If the ovaries are removed before three months, abortion occurs.

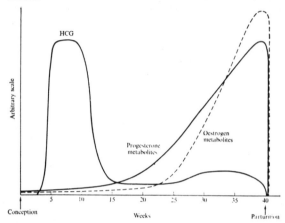

Figure 11.8 The urinary outputs of human chorionic gonadotrophin (HCG) and of oestrogen and progesterone metabolites during pregnancy.

The placenta is in fact an extremely important endocrine gland. The quantities of oestrogen and progesterone which it secretes increase steadily throughout pregnancy. In addition to these and to HCG it secretes two other hormones, making five in all:

1. *Human placental lactogen (HPL)*. This has some of the properties of growth hormone and some of prolactin. It may be important in stimulating breast growth, in regulating metabolism, in relaxing smooth muscle and in regulating the renal excretion of sodium, potassium and water. It may be the hormone responsible for the retention in the body of salt and water which occurs during pregnancy.

2. *Relaxin*. This relaxes the ligaments which hold the pelvic joints and softens the soft tissues of the birth canal, so making it easier for the baby to pass along.

MATERNAL PHYSIOLOGY

During pregnancy many changes occur in the maternal organism which enable the mother to maintain the pregnancy and to be prepared for the tests of labour and lactation.

Cardiovascular System

The blood volume increases by about 30% but the total red cell mass usually increases by less than this. As a result the red cell count and haemoglobin concentration tend to fall and this is so common that it has been called a 'physiological anaemia'. However, it has recently been shown that the anaemia may be partly due to folic acid and iron deficiencies since if adequate supplements of both are given the increase in red cell mass tends to keep pace with the increase in blood volume. The cardiac output also increases to about 30–40% above normal and does not fall until the baby has been delivered.

Smooth muscle in both arterioles and veins relaxes, possibly due to the combined effects of progesterone and HPL. There is a tendency for varicose veins to develop or to become worse partly because of smooth muscle relaxation and partly because the enlarging uterus impedes the return of blood from the lower limbs so tending to raise venous pressure.

Towards the end of pregnancy many women tend to faint if they lie on their backs. This is because the uterus presses on the inferior vena cava so preventing an adequate venous return of blood to the heart. If the blood cannot return it cannot be pumped out and so both cardiac output and blood pressure fall leading to fainting. This does not happen if a woman lies on her side.

Respiration
Breathing is stimulated, possibly by a direct action of progesterone on the brain. As a result the carbon dioxide level in the blood falls. The blood tends to become more alkaline and tetany is not uncommon.

Alimentary Tract
There is a general loss of smooth muscle tone which not uncommonly leads to constipation. In the early months of pregnancy there is often vomiting on getting up in the morning. The cause is unknown although it has been suggested that smooth muscle relaxation allows stomach fluid to pass back into the oesophagus where it leads to vomiting.

Urinary System
The blood flow to the kidneys and the glomerular filtration rate rise in parallel to the cardiac output. Many normal pregnant women have glucose in their urine, possibly because the increased filtration rate exceeds the reabsorptive capacity of the tubules. This finding does not mean that the woman is diabetic but may indicate the need for a glucose tolerance test.

The ureters share in the general loss of smooth muscle tone and become flaccid and dilated. The sphincters guarding the entrance to the bladder may become ineffective so allowing urine (and infection) to pass backwards from the bladder up towards the kidneys.

Hormones
The blood levels of thyroid, parathyroid and adrenal cortical hormones all rise. Parathyroid hormone mobilises bone calcium for transfer to the fetus. The high adrenal steroid levels may be responsible for the pigmented striae which often appear on the stretched abdomen and may also cause the remission of rheumatoid arthritis which sometimes occurs in pregnancy.

Weight
During a normal pregnancy the weight rises by 15–30 lb. About half of the rise can be attributed to the enlarging fetus, placenta and amniotic fluid volume. The other half is associated with increases in breast size, in blood volume, in interstitial fluid volume and in the deposition of body fat. There are very wide individual variations in the amount of weight put on and there is considerable argument about the mechanisms and meaning of the weight rise. Increased appetite may possibly be caused by progesterone and increased fluid retention by placental lactogen.

Pregnancy Tests
These all depend on the fact that very early in pregnancy large amounts of HCG are excreted in the urine. HCG can cause ovulation in many species of animal and most of the tests used until relatively recently depended on the effects of injecting suspected pregnancy urine into mice, rabbits or toads. These biological tests have now been superseded by exceedingly efficient biochemical tests which do not use animals.

PARTURITION
During pregnancy there is an enormous growth of the uterus. Within 40 weeks it develops from a small organ deep within the pelvis to an enormous one which alters all intraabdominal anatomy. Even in the non-pregnant state the uterus shows spontaneous contractions and this tendency of its muscle to contract is increased in pregnancy by two factors:

1. Stretching any form of smooth muscle tends to make it respond by contracting: uterine smooth muscle is no exception to this rule.

2. Oestrogens which are present in increasing concentrations as pregnancy progresses increase the power and excitability of the muscle.

What then prevents the uterus from contracting and expelling the infant long before the end of pregnancy? The answer is unknown but two factors may be involved. Both progesterone and placental lactogen may inhibit uterine muscle activity. The levels of lactogen tend to fall during the last two weeks before parturition and progesterone levels also fall immediately before and during labour. This then leaves the factors which stimulate uterine muscle action with a free field.

During the course of pregnancy uterine activity goes through the following phases:

1. In early and mid-pregnancy it is very much depressed, even as compared to the non-pregnant level.

2. At about the time of the 30th week, spontaneous contractions begin and can be recorded by special techniques. These contractions do not involve a coordinated response to the whole uterine muscle and they are not painful.

3. During the last two weeks or so of the pregnancy, the mother becomes aware of irregular uterine contractions which gradually become stronger and more painful.

4. True labour begins when the contractions become co-ordinated and regular, recurring at 10–15 minute intervals. Normally the contractions start near the top (fundus) of the uterus and then sweep down and around the uterus pressing the infant's head against the cervix.

5. The fetal membranes rupture releasing the amniotic fluid which they contain and allowing the infant's head to press harder on the cervix. This pressure on the cervix-activates sensory receptors: these receptors initiate a reflex which causes the release from the posterior pituitary gland of a hormone known as oxytocin which is a powerful uterine stimulant. The contractions become more and more powerful and closer and closer together in time so dilating the cervix and pushing the infant through and down into the vagina. At this stage the mother usually feels an irresistable urge to push downwards and to force the infant out.

6. Once the infant has been born, the placenta becomes detached from the wall of the uterus and is expelled: the uterus then contracts down over the raw surface to stop bleeding. Normally the expulsion of the placenta takes 30–60 minutes after birth of the infant but in most hospitals today delivery of the placenta is accelerated by the injection at the time of delivery of the infant of a drug which strongly stimulates uterine contractions.

The mechanism which starts off, the whole process of labour is unknown but recent work has suggested that the fetal pituitary and adrenal are important. In human infants or in animals whose pituitaries or adrenals are absent for some reason, pregnancy is almost always unduly prolonged. In experiments with sheep it has been demonstrated that this type of prolonged pregnancy can be quickly terminated by injecting cortisol through the abdominal wall into the *fetus*. Injection of cortisol into the mother has no effect. This suggests the the fetal pituitary at the appropriate time stimulates the fetal adrenals to secrete cortisol or a related hormone which then is an unknown way triggers off labour. A great deal of research remains to be carried out on this topic.

LACTATION

The process of lactation has three components:

1. The growth of the breasts from the non-pregnant state.

2. The synthesis and secretion of milk.

3. The ejection of the stored milk from the nipple.

The non-pregnant breast consist of some 15–25 lobes, each with a main duct which dilates and empties into collecting spaces beneath the pigmented area around the nipple (areola). Each main duct has many secondary ducts branching off it: these secondary ducts end in groups of secretory cells known as alveoli. Before pregnancy the ducts have few branches, the secretory cells are small and few in number and much of the breast is filled with fat. Oestrogen injections cause proliferation of the ducts while oestrogen and progesterone together stimulate growth of alveoli as well. Placental lactogen, thyroid hormone and adrenal steroids are also all necessary for normal breast growth during pregnancy.

The breasts become capable of secreting milk at any time after about the fourth month of pregnancy but virtually no milk is secreted until after parturition. This is probably because the process of secretion is blocked by the high levels of progesterone present during pregnancy. At birth progesterone levels in the blood fall sharply as the placenta is expelled. The levels of placental lactogen also fall but prolactin from the anterior pituitary takes over its function of breast stimulation. The first fluid to be

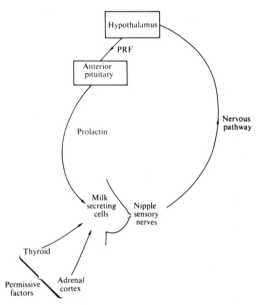

Figure 11.9 The control of milk secretion. Hormones from the thyroid and adrenal cortex (so-called 'permissive' factors) are essential for the normal secretion of milk although they do not take part in the reflex itself.

secreted is the watery, protein-rich, lipid-poor colostrum but within a few days normal milk is secreted at a high rate. The colostrum contains maternal antibodies which at this early stage in the infant's life can be absorbed from the gut and may help to give the child some immunity to disease during its early months.

The secretion of milk depends on a reflex in which both nerves and hormones are involved. Suckling stimulates the nerves in the nipple and impulses travel up to the hypothalamus. The hypothalamus then alters the output of a factor which travels down the pituitary portal blood vessels and controls the output of prolaction. So long as suckling continues, a high rate of prolactin secretion is maintained. When suckling stops the secretion of prolactin falls and the secretion of milk soon stops.

The ejection of the secreted milk from the breast also depends on the interaction of nerves and hormones.

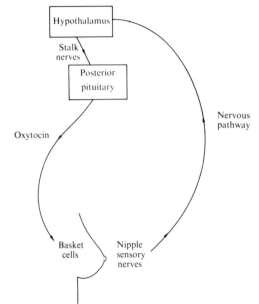

Figure 11.10 The control of milk ejection.

Again the process begins with suckling which activates nipple sensory nerves. Again the impulses ascend to the hypothalamus. But this time the hypothalamus sends impulses down to the posterior pituitary where oxytocin is released. The oxytocin then circulates in the blood back to the breasts. Surrounding the ducts of the breasts are special cells known as myo-epithelial cells which contract in response to oxytocin, so pushing the milk out. The hormonal part of the mechanism explains why the milk does not flow freely until a minute or two after suckling starts.

CONTRACEPTION

The only permanent answer to the population explosion lies in the increased availability and use of contraception. Broadly speaking, contraceptive techniques may be divided into those which require strong motivation, moderate intelligence and forethought before intercourse and those which do not require these factors. The main methods in use are as follows:

1. *Coitus interruptus*, or withdrawal of the penis just before male orgasm. This method is unsatisfactory and very frequently ineffective.

2. *The rhythm method*, which depends on the fact that in women with regular cycles ovulation rarely occurs outside the 9th and 19th days after the beginning of the menstrual bleed. Intercourse is avoided during the ten day interval. This method is ineffective in women who do not have regular cycles and requires very considerable self control by the partners.

3. *The rubber diaphragm* which fits into the vagina over the cervix and which is often combined with the use in the vagina of spermicidal jelly or foam. This method is effective when the diaphragm is fitted properly.

4. *The condom or sheath* which is fitted on to the erect penis and which catches the sperm and stops them entering the vagina.

5. *Oral contraceptives*, which seem to depend on the suppression of LH secretion and thus of ovulation by means of combinations of oestrogens and progestins. A pill is taken every day for three weeks and then withdrawn for a week during which an artificial, scanty menstrual bleed occurs. The method is 100% effective and is popular because it requires no thought or action before intercourse. It may cause depression, weight gain, high blood pressure and other changes in a relatively small proportion of users. In a tiny fraction of women it may cause clotting of the blood and any woman with a known tendency to abnormal clotting should not use the pill. In most countries the dangers of taking the pill are considerably less than the dangers of pregnancy.

6. *Progestin implants* which release a small amount of progestin daily and which can prevent conception for many months. The mode of action is uncertain but may be partly due to an action on ovulation and partly due to an effect on the uterus or Fallopian tubes.

7. *The intra-uterine device* (IUD or 'coil') which stops pregnancy by an unknown mechanism, possibly by interfering with implantation. A small plastic device is introduced into the uterine cavity and remains there until pregnancy is desired: it is only a little less effective than the pill. In a proportion of women it causes unpleasant side effects such as painful periods and it is rarely used in those who have not born children.

8. *Sterilisation*. Tying of the Fallopian tubes in women or the vasa in men leads to sterility. The female operation usually requires a general anaesthetic and abdominal incision while the male one can be done in a few minutes under local anaesthesia. The method is highly reliable but for obvious reasons is normally carried out only on those who have had at least two or three children. Considerable research is being carried out into techniques which will allow the operation to be reversible.

9. *Abortion*. This can be effective but surgical abortion can be carried out on a large scale only in highly developed countries where the hospitals, staff and finance are available. Surgical abortion used routinely as a method of controlling population leads to many ethical problems. It seems possible that in the future, however,

the place of abortion may be revolutionised by the widespread use of substances known as prostaglandins which were first isolated from prostatic fluid. They are very powerful stimulators of uterine muscle and in some centres the intravenous infusion of prostaglandin is used routinely for inducing abortion. The aim however is to produce either a pill or a vaginal pessary which contains prostaglandins and which can be used to stimulate uterine contractions and menstruation in any woman whose period is a few days overdue. The hope is that such medication may be able to be administered by the patient herself which would virtually eliminate the need for surgical abortion.

THE FETUS AND CHILD

The development of the fetus goes through two main stages. During the first ten weeks or so all the major organ systems of the body are formed. From about the 10th week to term all the organs grow and mature but there are few new major developments. Most important congenital abnormalities are determined by the 10th week of pregnancy and this is why it is particularly important to avoid exposure to rubella (German measles) and possibly harmful drugs during the first three months of pregnancy.

In a human being outside the uterus, lungs, gut and kidneys are all vital. But in the fetus the functions of these organs, providing food and oxygen and eliminating carbon dioxide and waste, are all carried out by the mother operating through the placenta. It is therefore not surprising that embryos with major defects in gut, lung and kidneys can survive until they are born at the normal time.

As well as allowing essential substances to pass easily, the placenta forms a very effective barrier to many factors which could harm the fetus. Bacteria cannot normally cross the placenta: the only major exception to this rule is syphilis and thus syphilis in the mother is almost always passed on to the child. Protozoa too do not normally cross the placenta, the main exception being toxoplasma which is very common throughout the world. In adults infection with toxoplasma (toxoplasmosis) is normally symptomless but in the fetus it may cause severe damage, especially to the brain. As might be expected from their small size, a number of viruses can cross the placenta. The best known of these is rubella but there are a number of others including the viruses which cause hepatitis.

The maternal and fetal circulations do not come into actual contact since they are separated by the trophoblast. However, during the violent uterine contractions which occur during parturition it is almost inevitable that some fetal blood enters the mother. This can of course lead to problems in later pregnancies if the mother is Rh− and the fetus is Rh+.

Another problem is that the fetus is immunologically different from the mother and therefore ought to be rejected by the mother like a foreign kidney or heart graft. Some theories of the beginning of labour have been based on the rejection idea. However it has now been demonstrated that the fetal trophoblast is covered by an immunologically inert layer, rich in sialic and hyaluronic acids. This layer keeps maternal and fetal tissues safely apart and prevents rejection.

FETAL CARDIOVASCULAR SYSTEM

In the fetus the cardiovascular system is the most important system of all since survival depends on the blood being pumped out along the umbilical cord to the placenta and back again. Fetal blood has a higher affinity for oxygen than adult blood: it contains a special type of haemoglobin, HbF, which has a very strong attraction for oxygen. HbF has virtually disappeared from the infant's blood within four months after birth.

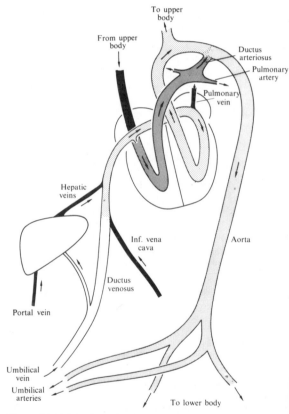

Figure 11.11 The fetal circulation. The degree of shading gives an indication of the degree of oxygenation, light indicating oxygenated and dark indicating deoxygenated blood.

The important features of the fetal circulation are shown in fig. 11.11. The main points to note are:

1. Oxygenated blood returns from the placenta via the umbilical vein. Most passes via the ductus venosus directly to the inferior vena cava where it is mixed with blood from the lower limbs. The rest goes to the left two

thirds of the liver and enters the vena cava via the hepatic veins. The right one third of the liver is supplied by the portal vein.

2. Two streams of blood enter the right atrium. These are a partially oxygenated stream from the inferior vena cava, and a deoxygenated stream from the superior vena cava. It has been shown that these two streams mix very little. The oxygenated blood flows straight through the foramen ovale into the left atrium while the deoxygenated stream passes to the right ventricle and is pumped out along the pulmonary artery.

3. Some of the pulmonary artery blood enters the pulmonary circulation and returns to the left atrium but most passes across into the descending aorta by way of the ductus arteriosus.

4. The oxygenated blood is pumped into the aorta by the left ventricle and enters the carotid and subclavian arteries.

5. Some of the blood in the descending aorta supplies the lower limbs and body but most passes via the paired umbilical arteries to the placenta.

THE SHOCK OF BIRTH
When the fetus emerges from the mother the umbilical cord is clamped and divided by the doctor or midwife so cutting off irrevocably the infant from the placenta. Even if the cord were not cut, on exposure to the outside air the umbilical vessels would contract and close as happens, of course, in animals. There is increasing evidence that immediate clamping of the cord before its natural pulsations have ceased may actually be harmful and may be a cause of anaemia in the new-born infant. Quite large amounts of fetal blood are present in the placenta and in the absence of interference uterine contractions force this blood along the umbilical cord into the infant. It is therefore sensible to wait for a few moments before clamping and tying the cord.

The immediate problem which faces the new-born child is that of supplying oxygen to its body. In the fetus, the lungs are functionless and they receive very little blood: most of the blood which enters the pulmonary artery is transferred directly to the aorta via the ductus arteriosus. In the new born infant, in contrast, the lungs are vital and must quickly expand while the pulmonary artery must carry to the lungs as much blood as the aorta carries to the rest of the body.

Respiration
In normal infants lung expansion presents few problems. Sensory stimuli resulting from the sudden emergence into the outside world initiate reflex breathing movements which automatically suck air into the chest. A few infants actually cry and breathe when only the head has emerged from the vagina and while the umbilical cord is still intact: in these cases only sensory stimuli from the head region must be required to initiate breathing. Other infants do not breathe until the cord has stopped pulsating or has been clamped and the oxygenation of the fetal blood falls. The least vigorous infants may give only ineffective gasps or may fail to breathe at all. Maternal anaesthesia and analgesia and long and difficult labours in which the cord may be compressed and the oxygen supply to the fetus reduced long before birth actually takes place are all associated with trouble in initiating breathing. In these cases slapping, cold or warm water on the skin, distension of the anal sphincter and (most

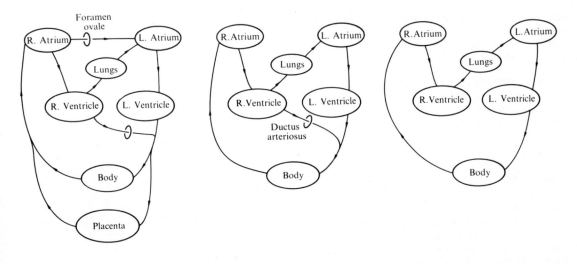

Foetal Neonatal Adult

Figure 11.12 Outlines of the circulation in the fetus, new-born infant and adult.

important of all) cautious artificial ventilation with an oxygen-rich mixture may start respiration. The force that must be applied to expand the lungs in the first breaths is very large and in the absence of alveolar detergent lung expansion may be impossible and the infant may die of the respiratory distress syndrome.

Circulation

Changes in the circulation are just as important as and are complementary to the changes in the lungs. The main features are:

1. When the umbilical cord closes down there is a sharp increase in the resistance to the outflow of the left ventricle. As a consequence, the blood pressure in the aorta, left ventricle and left atrium all tend to rise.

2. The closing of the cord cuts off the venous return from the placenta and reduces the inflow of blood to the right side of the heart. Pressures in the right atrium, right ventricle and pulmonary artery all fall.

3. Expansion of the lungs greatly reduces the resistance to blood flow through them. This lowering of pulmonary resistance tends to lower blood pressure in the pulmonary artery, right ventricle and right atrium.

4. Because of these changes, pressure in the right atrium becomes lower than in the left atrium. As a result the foramen ovale closes with a valve-like action and soon becomes permanently sealed.

5. Because the pressure in the aorta becomes higher than that in the pulmonary artery, the direction of blood flow through the ductus arteriosus changes. It flows from aorta to pulmonary artery instead of in the reverse direction. This means that the blood from the aorta goes round to the lungs again for a second helping of oxygen. When the oxygen saturation of the blood flowing through the ductus reaches about 90%, the ductus closes and the adult form of circulation becomes operative.

THE NEW BORN INFANT

Once the infant has overcome the immediate shock of adjusting its respiratory and cardiovascular system to the demands of the outside world it is faced with further problems.

The Gut

At birth the gut is filled with a substance known as meconium which consists of partially digested epithelial cells which have been shed from the gut wall and which is usually stained green with biliverdin, a breakdown product of bile. The stools are therefore at first green. Within a few days they become brownish-yellow. This is an important sign because it indicates that new faeces are being formed from milk products and that the gut must therefore be open along its whole length. Many congenital abnormalities can cause partial or complete blockage of the gut.

During the first few days of life all infants lose weight, partly because of inadequate water intake and partly because the nutritive value of maternal milk may be poor at this time. Most milk diets are deficient in iron and so until it is weaned the infant must survive on its own iron stores built up during the last part of intra-uterine life.

Temperature Regulation

Body temperature in infants is unstable. This is primarily because methods of controlling heat loss using the skin circulation and sweat glands are not well developed. The shivering mechanism is also ineffective in infants. It is therefore very important to keep new born infants warm.

Liver

The liver contains large stores of glycogen at birth but within a few hours the glycogen is used up by conversion to glucose and the blood glucose level falls. The fall in blood glucose is particularly severe in small and premature infants and if untreated it can cause brain damage or even death. It is therefore important to check on the blood glucose concentration in infants who are at risk.

In the fetus, bilirubin resulting from haemoglobin breakdown is for the most part disposed of across the placenta. After birth, before bilirubin can be excreted into the gut, it must be conjugated with glucuronic acid by the liver. During the first week of life the mechanisms for carrying out this reaction are poorly developed. Free bilirubin therefore tends to accumulate in the blood and may cause jaundice. This is unlikely to be dangerous in a normal infant but in premature babies and in those in whom haemolysis is excessive because of rhesus disease, dangerously high levels of bilirubin may occur and can cause brain damage (kernicterus).

CHILDHOOD

The most obvious physical features of childhood are growth in height and weight. The tissues of the body can

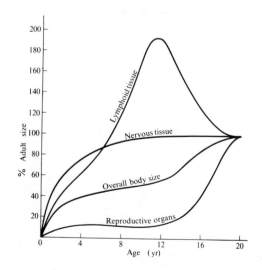

Figure 11.13 Patterns of growth in different tissues.

be divided into four groups according to the growth curve which they follow. With the exception of the reproductive

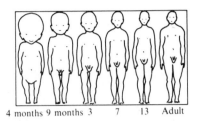

4 months 9 months 3   7   13   Adult

Figure 11.14 The changes in body shape and contour which take place during life.

organs, all tissues grow rapidly after birth. Rapid growth of nervous tissue continues until the age of 8–9 when it has almost reached the final adult mass. Lymphoid tissue has an unusual pattern reaching very high levels around puberty and then falling to the final adult mass during the teen years. The reproductive organs grow little until puberty when there is a sudden spurt. There are also changes in overall body shape. The decreasing relative size of the head is a reflection of the early growth of nervous tissue.

The maximal height to which an individual can grow is probably genetically determined but many other factors decide whether the maximal height will be reached. Nutrition is the most important of these and malnutrition is a major cause of poor physical development. Normal

growth seems to require the interaction of growth hormone from the pituitary, the adrenal hormones and thyroid hormone.

PUBERTY

Puberty is the period of life when the gonads grow rapidly, when the output of sex hormones leads to the development of secondary sexual characters and when mature ova or sperm are released for the first time. The main features of puberty were described in the last chapter. In girls the first signs are pubic hair growth and breast development: they precede menarche by 1–2 years. In boys growth of pubic hair and the penis are the first signs. In girls the first signs of puberty tend to occur 1–2 years earlier than in boys but there is wide individual variation. For example about 5% of normal girls show some sign of puberty at the age of 10 while about 5% of normal boys show no signs until they are 15.

Before puberty there is relatively little difference between the heights and weights of girls and boys. Boys tend to be a little taller during infancy and very early childhood while between the ages of 8 and 13 girls may become taller and heavier because they show an earlier pubertal growth spurt. This spurt is usually before the menarche and relatively little growth occurs afterwards. Boys eventually become taller and heavier than girls because they have a longer prepubertal growth period and the pubertal spurt is greater. Adrenal androgens are probably very important in causing growth at the time of puberty.

12 Responses of the Whole Body

Up until this chapter we have been primarily concerned with discussing the behaviour of individual systems in the body. This is clearly an oversimplified and artificial way of looking at things for in the normal individual all the systems work closely together for the good of the whole. In this chapter we shall be considering various situations in which the physiology must be considered in terms of the responses of the whole body.

TEMPERATURE REGULATION

Everyone knows that the 'body' has a 'steady temperature' but not many know precisely what the terms 'body' and 'steady temperature' mean. The parts of the body whose temperature is most closely regulated are the arterial blood and the brain. Yet even these tissues do not have a temperature which is absolutely steady at 98·6 °F or 37 °C. There is considerable variation during the day. In most people the daily variation in temperature is around 1 °F but in some entirely normal individuals it may be as much as 2 °F. Temperatures are lowest in the early morning and highest in the early evening, a normal variation of from 97 °F to 99 °F being not unknown. A minor degree of apparent temperature elevation early in the evening is therefore not necessarily pathological.

As well as variation of body temperature with time there is also variation between different parts of the body. Most of the deep organs closely reflect the temperature of arterial blood but working muscles and the highly active liver may be 1–2 °F higher than the blood. Skin temperature varies enormously. It is usually much colder than that of arterial blood but in hot climates when the skin blood vessels are open wide in order to lose heat the skin temperature may be close to that of arterial blood.

Arterial blood temperature is normally measured indirectly by thermometers placed in the mouth, axilla or rectum: the thermometer should usually be left in position for two minutes if an accurate reading is to be obtained. Of the three sites the temperature of the mouth most closely reflects that of arterial blood. The tempera-ture of the axilla tends to be a little lower and that of the rectum a little higher. With infants it is easiest to take their temperatures by means of a rectal thermometer.

The reason for the need to maintain a relatively steady body temperature is that many of the basic biochemical and physical processes which go on in the body are temperature dependent. In general, heat tends to speed them up and cold to slow them down, although if the heat is sufficient to damage protein molecules then heat may cause destruction of cells. The behaviour of enzymes and of the nervous system is particularly temperature dependent. One of the main features of either too high a temperature or too low a temperature is disordered brain function with the delirium of fever or the sluggishness and coma of hypothermia (low temperature).

BASAL METABOLIC RATE (BMR)

The metabolic rate is defined as the number of calories used in a given time. The basal metabolic rate (BMR) is the metabolic rate of a subject at rest who has not had any food for at least 12 hours. The BMR therefore represents the amount of energy required to keep the body running when no external work is being done and no food is being digested.

The BMR of a healthy adult man might be in the region of 2,000 kcal/day. The BMR of a healthy child might be 1,000 kcal/day while the BMR of a mouse might be 4 kcal/day. This does not necessarily indicate any fundamental differences in metabolism but merely means that more calories are required to run a larger body. If BMR is expressed as kcal/kg/day, the differences are much reduced but are still present. However, if BMR is expressed as kcal/square metre of body surface/day, all mammals are found to be remarkably similar. The BMR is therefore very closely related to the surface area. This may be because heat is lost from the body surface and so the amount of heat which must be generated to replace the loss must be related to the surface area.

Measurement of the BMR

The most accurate way of measuring the BMR is to measure the total heat output of an individual by enclosing him in a huge calorimeter. However such direct calorimetry is laborious and very expensive and an indirect method is used for routine measurements of metabolic rate.

Energy is obtained in a useful form by the oxidation of foodstuffs. Except in the presence of an oxygen debt (see later this chapter) the amount of energy released is proportional to the amount of oxygen used up. If pure protein is being oxidised, when 1 litre of oxygen (at STP) is used up, about 4·5 kcal are made available. The corresponding figure for fat is 4·7 kcal/L and for carbohydrate is 5·0 kcal/L. The mixture of the three which is oxidised by normal humans under basal conditions yields about 4·8 kcal/L. Using this figure, if the oxygen consumption over a given period is known, then the metabolic rate for that period can be easily calculated. For instance, if 20 litres of oxygen were consumed in an hour, the metabolic rate would be roughly $20 \times 4·8 = 96$ kcal/hour. The oxygen consumption is usually measured by an instrument known as a spirometer which measures the rate at which oxygen is used for a chamber of known volume.

CONTROL OF HEAT PRODUCTION

As a result of its metabolic activities the body continually produces heat. There are four main ways in which the amount of heat produced by the body may be increased:

1. By voluntary exercise.

2. By the form of involuntary exercise known as shivering.

3. By increasing the rate of secretion of adrenaline (epinephrine) from the adrenal medulla. This breaks down glycogen in liver and muscle and increases the metabolic rate by increasing the availability of glucose.

4. By increasing the output of thyroid hormone which raises the basal metabolic rate. This seems to be a long term response which occurs in those exposed to cold for prolonged periods but it does not seem to operate in the short term.

CONTROL OF HEAT LOSS

The rate of transfer of heat from the body to the outside world depends on the following factors:

1. The temperature gradient between the body and its surroundings. If the environmental temperature is very low, heat will be lost rapidly. In a few parts of the world during the day time the environmental temperature may be above body temperature and so the body may actually gain heat from its surroundings.

2. The amount of body surface exposed and the effectiveness of the insulation of the body coverings. Heat is lost or gained most easily across bare skin and the amount of heat transfer decreases as the effectiveness of the insulation provided by clothing increases.

The main ways in which the body can regulate the rate of heat loss are as follows:

1. *By changing behaviour.* Heat loss will clearly be more effective in a cool environment than in a hot one. It is therefore possible to increase the rate of heat loss by moving to a cool place and seeking shade. Only 'mad dogs and Englishmen go out in the mid-day sun'.

2. *By altering the amount of clothing worn.*

3. *By altering the amount of blood flowing through the skin.* Skin blood flow is extremely variable. When all the skin vessels are opened up the skin is as warm as arterial blood and heat loss from the body will be made easier. When all the skin vessels are closed down, skin temperature will become similar to that of the environment and there will therefore be minimal heat transfer between the skin and its surroundings. When the skin vessels are closed down, there is then a layer of insulating fat between the deep parts of the body and skin.

4. *By altering the rate of sweating.* Sweating is an effective way of losing heat because when 1 ml of water evaporates, it takes about 540 calories from its surroundings. This heat is known as latent heat and it means that when water evaporates from the skin, the skin will be cooled. Evaporation and therefore heat loss can occur even in situations when the surrounding environment is hotter than the body itself. However, the ease with which water evaporates depends on the relative humidity of the air. If the relative humidity is 100% this means that the air is carrying all the water vapour that it can and no more can evaporate. Sweating will therefore be an ineffective method of heat loss and the sweat will run off the skin without evaporating. This explains why hot, humid, 'sticky' environments such as the West Coast of Africa or the Mississippi Valley of the USA are so uncomfortable. In contrast, if the relative humidity is low, as in deserts, evaporation is rapid and sweating is a very effective way of losing heat. This is why desert climates are often relatively pleasant even though the actual environmental temperature may be very high.

THE TEMPERATURE CONTROL SYSTEM

Like all control systems, the one which regulates temperature has three basic components, receptors, a central control mechanism and effectors. We have already discussed the effectors, the mechanisms for regulating heat production and heat loss. Where are the receptors and the control centre? There are two sets of receptors:

1. *Receptors in the skin* which measure the temperature of the environment. Everyone is aware of the functioning

of these receptors. They are important in producing modifications of behaviour and in warning the control centre of changes in the environment which may lead to changes in the temperature of the body.

2. *Receptors in the brain* (hypothalamus) which measure the temperature of the arterial blood itself.

The central control mechanism is also in the hypothalamus. It receives information from all the receptors and in accordance with this it modifies the behaviour of the effectors. If the body temperature is too high, heat production is cut down by reducing muscular activity and heat loss is increased by seeking a cool environment, by sweating and by increasing the blood flow to the skin. If the body temperature is too low, heat production may be increased by exercise, shivering or adrenaline while heat loss may be cut down by reducing skin blood flow. The skin, especially in the extremities, becomes cold and often blue due to deoxygenation of the sluggishly flowing blood.

FEVER

The commonest abnormality of temperature regulation is fever, the elevation of temperature which occurs in inflammatory disease of any sort whether due to infection, tissue damage because of injury, or cancer. The function of fever is unknown: it does not appear to enhance the body's ability to cope with disease. The only common organism which may be killed by the temperatures reached in fever is the spirochaete of syphilis. Unfortunately syphilis itself does not produce fever of sufficient degree to kill the spirochaetes. However, if someone with syphilis becomes infected with malaria, the high temperatures reached because of the malaria will kill the spirochaetes. Before the development of modern chemotherapy it was therefore not uncommon to treat syphilis by causing a deliberate malarial infection.

In fever the central regulating mechanism is reset so that it operates around a new temperature. Instead of regarding, say, 98·6 °F as the normal body temperature, the control centre comes to regard, say 104 °F, as 'normal'. A truly normal temperature is therefore far too low as far as the hypothalamic control centre is concerned. The control centre therefore orders an increase in heat production by shivering and a reduction in heat loss by means of constriction of skin blood vessels: the patient feels cold. As a result of these effector actions the body temperature rises until it reaches 104 °F and the hypothalamus regulates around this new level. At the termination of the fever the hypothalamus again comes to regard a temperature of, say 98·6 °F, as normal. 104 °F is obviously far too high and the temperature is rapidly brought down by sweating and an increase in blood flow to the skin.

The change in behaviour of the hypothalamus in fever seems to be brought about by substances called pyrogens which are high molecular weight carbohydrates. Pyrogens are released by leucocytes which become involved in an inflammatory reaction and circulate in the blood to act on the brain.

MENSTRUAL CYCLE

Women show variations of the early morning arterial temperature during the menstrual cycle. Body temperature is at a minimum at the time of menstruation, rises slowly during the first half of the cycle and then shows a sharp jump of about 0·5 °C at the time of ovulation. The increase is probably due to action on metabolism of progesterone which is secreted at about this time. The rise in temperature can be used to indicate ovulation both in women desiring conception and in those trying to avoid it.

HEAT STRESS

People who live in hot environments may show a number of different types of illness associated with the problem of keeping cool.

1. *Cramps.* Those who work in a hot and humid environment often suffer cramps. These occur because sweat contains salt as well as water and so at high rates of sweating there is considerable salt loss. Loss of sweat tends to be replaced by the drinking of water only: the body therefore becomes salt depleted and cramps result. The condition may be cured by drinking dilute salt solutions for fluid replacement or by taking ample salt with meals.

2. *Prickly heat.* During periods of prolonged sweating, the superficial layers of the skin may become soft and block the ducts of the sweat glands. Sweat therefore cannot escape and produces little pustules which are extremely itchy and very easily become infected.

3. *Heat exhaustion and heat stroke.* These occur when fluid is lost by sweating and not replaced. First there is a feeling of light headedness and weakness which may progress to weakness, nausea and vomiting. Up to this stage body temperature is usually little above normal and the patient recovers quickly if given adequate fluids and salt by mouth: this stage is often known as heat exhaustion. However, if the sweat loss continues further without replacement a stage may eventually be reached when the body water is so depleted that sweating stops and the blood becomes very viscous because of the water loss. When sweating stops, the body temperature begins to rise rapidly and if nothing is done death will occur soon, preceded by delirium and then unconsciousness. This stage is known as heat stroke and requires urgent treatment. The patient must be cooled as quickly as possible by being bathed in iced water. At the same time fluids and electrolytes must be replaced intravenously and the limbs must be vigorously massaged to increase skin blood flow and heat loss.

EXERCISE

Exercise is another excellent example of the interaction of several systems of the body to produce a coordinated response. Exercise begins when the central nervous system sends instructions to the muscles to begin working. The muscles move the joints and a whole complex series of events is initiated.

RESPIRATION

The control of ventilation during exercise can be divided into two phases, depending on whether the ventilation rate is below or above about 30 L/min. Below a rate of about 30 L/min the control of respiration seems to depend on the combined information provided by carbon dioxide and joint receptors and has been discussed in chapter 8. The outstanding features of exercise at this level are that the pCO_2 of arterial blood remains constant and that breathing 100% oxygen does not reduce the ventilation rate, indicating that oxygen lack is not playing a significant role. The constancy of the pCO_2 suggests that this is the dominant factor.

With more severe exercise, the pCO_2 of arterial blood surprisingly actually falls, indicating that the ventilation rate is more than sufficient to maintain a normal pCO_2. This suggests that other factors are now dominating the control system. The most important ones probably are:

1. A stimulus of oxygen lack. At this level breathing 100% oxygen reduces the ventilation rate.

2. A fall in blood pH due to the release of acids such as lactic acid from the exercising muscles. The change in pH stimulates the peripheral chemoreceptors.

3. A rise in body temperature occurs and this also increases ventilation possibly by directly affecting the respiratory control centre.

As a result of all these stimuli the ventilation rate during very severe exercise may approach 100 L/min. At this level, however, increases in ventilation rate have little effect in increasing the available oxygen supply. This is because virtually all the additional oxygen is used up by the additional activity of the respiratory muscles themselves.

CARDIOVASCULAR SYSTEM

It would not be of much use to supply more oxygen by increasing the ventilation rate, if the cardiac output were not increased to supply the oxygen rapidly to the working muscles. The cardiac output may rise from 4–5 L/min at rest to over 35 L/min in very severe exercise. The rise in cardiac output is brought about partly by an increase in heart rate and partly by an increase in stroke volume. The heart muscle contracts more vigorously and with each beat pumps out a greater proportion of the blood it contains. Sympathetic nerves to the heart and adrenaline released from the adrenal medulla are important in stimulating the heart to beat more rapidly and more vigorously.

Not only does the cardiac output increase but the distribution of blood to the various organs changes. Arterioles in the muscles open wide partly because of reduction of sympathetic activity but more because of the local

effects of high carbon dioxide and reduced oxygen levels. The dilatation results in a large increase in muscle blood flow. Part of the demand is met by the increase in cardiac output and part by a reduction of flow to the gut and kidneys to about one fifth of the normal rate or even less. This is why it is unwise to do strenuous exercise immediately after a meal: blood is diverted from the gut and digestion does not take place satisfactorily.

The blood flow to the skin is important in exercise because it offers a way of getting rid of the extra heat produced by muscular activity. The skin blood vessels open wide giving a warm, red skin. Sweating is usually profuse. If the rate of heat elimination were not greatly increased during exercise the body temperature could rise rapidly and dangerously: as it is a rise of 2–3 °F is not uncommon.

METABOLISM

One of the major problems in exercise is to provide fuel for energy to the working muscles while simultaneously not allowing the muscles to take so much glucose from the blood that the energy supply to the brain is put in danger. This is achieved primarily by the actions of adrenaline. Adrenaline acts in three main ways:

1. It breaks down glycogen to glucose in the liver. This glucose enters the blood.

2. It reduces the rate at which muscle removes glucose from the blood, so conserving glucose for the brain.

3. Within the muscle fibres it stimulates the breakdown of glycogen to glucose so increasing the internal supply of glucose to the muscle.

During exercise there is a markedly increased rate of oxygen consumption but afterwards this does not immediately return to its original level. It remains high for a period which after severe exercise may be as long as an hour. This raised post-exercise consumption is said to be the paying off of an 'oxygen debt' by the body. The debt may be incurred in two ways:

1. At the beginning of exercise, oxygen may be consumed faster than it can be supplied by the pulmonary and cardiovascular systems. This applies particularly to brief periods of intense exercise such as a 100 m dash which may be performed with the runners hardly taking a breath.

2. Even in steady state exercise when the rate of oxygen consumption is equal to the rate of oxygen provision, muscle metabolism may not be fully aerobic. Large amounts of ATP may be provided by the simple anaerobic conversion of glucose to lactic acid. After the exercise lactic acid still in the muscles is reconverted to pyruvate.

Lactic acid in the blood may diffuse either into muscle or into liver where it is also converted to pyruvate and built up into glycogen. The disposal of lactic acid and of other metabolites which may accumulate during exercise requires oxygen and so the rate of oxygen consumption will be above resting levels.

ACID AND BASES

Before one can understand the mechanism of acid-base balance it is essential to have an elementary but nevertheless very clear understanding of the fundamental properties of acids and bases. An acid may be defined as a substance which can add hydrogen ions to solution. A base is a substance which can remove hydrogen ions from solution. The acidity of a solution depends on its hydrogen ion concentration. Strong acids are substances such as hydrochloric acid which donate *all* their hydrogen ions to solution and which effectively do not exist in molecular form. Weak acids are substances like fatty acids and amino acids which donate only some of their potential hydrogen ions to solution and which exist partly in molecular and partly in ionic form.

The equilibrium governing the ionisation of water is as follows:

$$H^+ + OH^- \rightleftharpoons H_2O$$

By the Law of Mass Action,

$$\frac{[H^+]\ [OH^-]}{[H_2O]} = K$$

where K is a constant. However, the amount of undissociated water is enormous relative to the amount of the charged particles and can usually be regarded as constant. The equation can therefore be simplified to read

$$[H^+]\ [OH^-] = Kw$$

The numerical value of Kw in the case of pure water is 10^{-14}. Since in pure water the concentrations of hydrogen and hydroxyl ions are equal, the concentration of each must be 10^{-7} mEq/L. The acidity of a solution depends on its hydrogen ion concentration but to state

Table 12.1 Hydrogen ion concentration expressed in three different ways.

1.0N	1 N	pH 0
0.1N	10^{-1} N	pH 1
0.01N	10^{-2} N	pH 2
0.001N	10^{-3} N	pH 3
0.0001N	10^{-4} N	pH 4
0.00001N	10^{-5} N	pH 5
0.000001N	10^{-6} N	pH 6
0.0000001N	10^{-7} N	pH 7
0.00000001N	10^{-8} N	pH 8
0.000000001N	10^{-9} N	pH 9
0.0000000001N	10^{-10} N	pH 10
0.00000000001N	10^{-11} N	pH 11
0.000000000001N	10^{-12} N	pH 12
0.0000000000001N	10^{-13} N	pH 13
0.00000000000001N	10^{-14} N	pH 14

the concentration in the form just used is rather clumsy. Sorensen introduced the simple pH scale. He took the negative logarithm to the base 10 of the concentration and called the result pH. For the mathematically vague, all this means is that he took the exponent (the little number above and to the right of the 10) and made it positive. Pure water therefore has a pH of seven on this scale. In order to clarify the meaning of pH, Table 12.1 describes the acidity of a solution in three different ways. Two points in particular should be noted.

1. A change of one unit on the pH scale means a tenfold change in hydrogen ion concentration. A twofold change in hydrogen ion concentration is expressed by a movement of 0.3 on the scale.

2. The lower the number on the pH scale, the greater the acidity. A falling pH means a rising acidity and vice versa.

Consider now the equilibrium which holds in a pure solution of a weak acid. $\quad HA \rightleftharpoons H^+ + A^-$

By the Law of Mass Action,

$$K = \frac{[H^+]\ [A^-]}{[HA]}, \text{ or } [H^+] = \frac{K[HA]}{[A^-]}$$

Take logarithms of both sides of the equation (if you have forgotten your algebra, it may help you to remember that when logarithms are taken, expressions which were originally divided are subtracted).

$$\log [H^+] = \log K + \log [HA] - \log [A^-]$$

Multiplied by -1,
$$-\log [H^+] = -\log K - \log [HA] + \log [A^-]$$

But Sorensen defined $-\log [H^+]$ as pH and, by similar reasoning, $-\log K$ can be written pK. The expression therefore becomes

$$pH = pK + \log \frac{[A^-]}{[HA]}$$

But when half the molecules of the weak acid HA have dissociated, $[A^-] = [HA]$ and $\log [A^-]/[HA] = \log 1 = 0$. Therefore in this situation, pH = pK. This means that the pK of a weak acid is the pH at which half the molecules of the weak acid are dissociated. Precisely similar reasoning can be applied to weak bases. The following equation is an example

$$-NH_3^+ = -NH_2 + H^+$$

In a practical situation, it us very rare to find a pure weak acid or a pure weak base in solution. It is much more common to find a combination of the weak acid or base with a salt of that acid or base. For instance, in plasma, carbonic acid is found in conjunction with its sodium salt, sodium bicarbonate. In general terms, we may say that four types of particles exist in such a solution, the undissociated acid molecules (HA), the ions resulting from dissociation of the acid (H^+), the negative ions arising from dissociation of the acid and from the salt (A^-) and the positive metallic ions of the salt (B^+). The salt is completely dissociated and its contribution to the total concentration of A^- will be enormous relative to the contribution coming from the dissociation of HA. Therefore as an approximation we may say that the concentration A is equal to the salt concentration. Modifying equation (1) we may therefore write;

$$pH = pK + \log \frac{[\text{salt}]}{[\text{acid}]}$$

This modification is usually known as the Henderson–Hasselbach equation.

BUFFERS

The importance of systems of weak acids or weak bases lies in their power of buffering. Suppose we start with a solution of a weak acid and we add enough hydrogen ions to make the solution very acid. These hydrogen ions will combine with all the available A^- ions and so all the acid will be in the form HA. Suppose now we begin to remove hydrogen ions by adding a base which mops them up and that we plot on a graph the change in pH for each addition of base. Suppose the solution is at pH 1 to start with and that enough base is added to change it to pH 2, i.e. to lower the hydrogen ion concentration to one tenth of its original value. Suppose then we again add enough base to remove nine-tenths of the remaining hydrogen ions. We find that the pH does not fall to 3 as expected. The base has removed the expected number of hydrogen ions, but some of these have been replaced by hydrogen ions arising from the breakdown of HA. As more base is added, more HA dissociates and far more base than expected is required to bring about a given change in pH. Eventually, however, all the HA molecules are used up, the resistance of the solution to change in pH diminishes and once more a given amount of base produces the expected change in pH. The change in pH on addition of base is shown in the titration curve in fig. 12.1. At the mid-point of the steep part of the curve, the

concentration of HA equals the concentration of A^- ions. At this point, therefore, the pH of the solution equals the pK of the acid. The solution at this point resists equally well changes in either direction. The buffering powers of a weak acid or weak base system are therefore maximum at the pK value. If a substance has more than one weak acidic or basis group, each group has its own characteristic pK and optimum buffering region.

The most important buffering reactions within the body are shown in fig. 12.2. All the substances on the

$$H_2CO_3 \rightleftharpoons H^+ + HCO_3^-$$

$$HbH \rightleftharpoons H^+ + Hb-$$

$$H_2PO_4^- \rightleftharpoons H^+ + HPO_4^=$$

$$-COOH \rightleftharpoons H^+ + -COO-$$

$$-NH_3^+ \rightleftharpoons H^+ + -NH_2$$

$$NH_4^+ \rightleftharpoons H^+ + NH_3$$

Figure 12.2 The most important buffering reactions in the body.

right hand sides of the equations can combine with hydrogen ions and remove them from solution and are therefore bases. All those on the left hand side of the equation can donate hydrogen ions to solution and are therefore weak acids.

ACID-BASE BALANCE

The most important single reaction in the body as far as acid-base balance is concerned is the top one in fig. 12.2. If we substitute carbonic acid and bicarbonate ions in the Henderson-Hasselbach equation it becomes

$$pH = pK + \log \frac{[HCO_3^-]}{[H_2CO_3]}$$

In practice the concentration of carbonic acid is impossible to measure directly but it is proportional to the pCO_2 i.e. the carbonic acid concentration is equal to the pCO_2 multiplied by a constant. This means that the equation can be written:

$$pH = pK + \log \frac{[HCO_3^-]}{K_1 \cdot pCO_2}$$

It is obvious that in this equation there are only three variable parameters, the pH, the bicarbonate concentration and the pCO_2. In an equation containing three variables, if two of them are determined the third is automatically also determined. The acid-base balance system of the body operates because the respiratory system maintains the arterial pCO_2 constant and the kidneys maintain

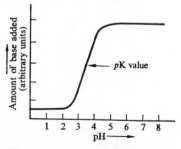

Figure 12.1 The titration curve of a weak acid.

arterial bicarbonate concentration constant. The pH is therefore automatically also maintained constant.

The regulation of pCO_2 by the respiratory system has been explained in chapter 8. The regulation of the plasma bicarbonate concentration by the kidneys is closely bound up with the secretion of acid by the renal tubules. The secretion of hydrogen ions by the tubules (chapter 9) depends on the supply of hydrogen ions available within the cells. The hydrogen ion concentration within the cells is determined partly by the pH of the plasma and partly by the pCO_2 within the cells: in turn the intracellular pCO_2 depends on the plasma pCO_2. In a normal individual, with an arterial pH of 7·4 and an arterial pCO_2 of about 40 mmHg, the kidney tubules are capable of secreting about 3 milliequivalents of hydrogen ions per minute. Also in a normal individual with a glomerular filtration rate of 125 ml/min and a plasma bicarbonate concentration of 24 mEq/L, 3 milliequivalents of bicarbonate will reach the tubules each minute. Fig. 12.3

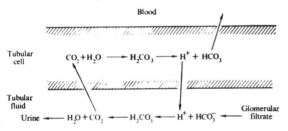

Figure 12.3 The renal tubular mechanisms for the secretion of hydrogen ions and the reabsorption of bicarbonate.

shows that for every hydrogen ion secreted into the urine, one bicarbonate ion is secreted into the blood. If bicarbonate is present in the urine, for every hydrogen ion secreted, one bicarbonate ion is reabsorbed. Thus at an arterial pH of 7·4, an arterial pCO_2 of 40 mmHg, and an arterial bicarbonate concentration of 24 mEq/L, the bicarbonate which is filtered will just balance the hydrogen ions which are secreted. Urine pH will be neutral.

What will happen if the hydrogen ion concentration rises because of a change in metabolic activity or because of ingestion of hydrogen ions from the gut? Initially, the top equation in fig. 12.2 shows that there must be a slight rise in hydrogen ion concentration coupled with a slight fall in bicarbonate concentration. The fall in pH will raise the rate of hydrogen ion secretion by the kidneys and the fall in bicarbonate concentration will diminish the rate of bicarbonate filtration. There will therefore now be an imbalance between hydrogen ion secretion and bicarbonate filtration. As a result some hydrogen ions secreted will not be buffered by bicarbonate and will not be reabsorbed in the form of carbon dioxide. They will be buffered by phosphate or ammonia and will tend to make the urine acid. These hydrogen ions will therefore be lost to the body. At the same time, because of the mechanism

in fig. 12.3, for every hydrogen ion secreted, one bicarbonate ion will be added to the blood. This means that during its passage through the kidneys the blood will receive more bicarbonate from the tubule than it loses by glomerular filtration. The net effect therefore will be to raise the plasma bicarbonate concentration and lower the hydrogen ion concentration until the rate of hydrogen ion secretion is again equal to the rate of bicarbonate filtration, thus once more establishing equilibrium.

The opposite sequence of events will occur if there is a rise in plasma bicarbonate concentration. This will lead to a fall in plasma hydrogen ion concentration and a fall in the rate of tubular secretion of hydrogen ions. The rate of bicarbonate filtration will therefore exceed the rate of hydrogen ion secretion and hence of bicarbonate reabsorption. Bicarbonate will be lost in the urine and hydrogen ions will be retained in the plasma until equilibrium is again established between hydrogen ion secretion and bicarbonate filtration.

Thus by this mechanism, although there are individual differences, the bicarbonate concentration of the plasma is maintained by the kidney at around the 24 mEq/L mark.

ACID-BASE PROBLEMS

There are two main problems in the regulation of acid-base balance:

1. Considerable amounts of carbon dioxide enter the blood in the capillaries. When the body is in equilibrium, equal amounts of carbon dioxide are removed again in the lungs. However, the amount of carbon dioxide added could potentially produce a large drop in the pH of capillary and venous blood and it is desirable that this impact should be minimised. This is achieved because the buffer systems of the blood remove from solution many of the hydrogen ions which are added by the carbon dioxide. Haemoglobin is by far the most important of these buffers. When haemoglobin gives up oxygen, it becomes a weaker acid than when it is fully oxygenated. Weak acids take up hydrogen ions and for every 10 millimoles of oxygen given off, 7 milliequivalents of hydrogen ions are removed from solution by the haemoglobin. This is almost sufficient by itself to take up all the hydrogen ions produced by the addition of carbon dioxide to the blood. The remaining hydrogen ions are buffered by plasma proteins and other buffers and the actual pH change is very small.

2. When the body is in equilibrium, the kidneys and lungs eliminate each minute precisely the same number of hydrogen ions as is formed. This equilibrium can be upset in four major ways which are outlined in the remainder of this section.

RESPIRATORY ACIDOSIS

This occurs when because of some disease the respiratory system cannot eliminate carbon dioxide at the normal rate and the pCO_2 of arterial blood rises. Equation (1) shows that if the pCO_2 rises the pH must fall. The increased pCO_2 and the rise in hydrogen ion concentration enable the kidneys to secrete more hydrogen ions. Each additional hydrogen ion secreted into the urine is accompanied by a bicarbonate ion secreted into the blood. The plasma bicarbonate concentration therefore rises until it balances the raised pCO_2 and arterial pH returns to normal.

RESPIRATORY ALKALOSIS

This occurs when the lungs get rid of carbon dioxide more rapidly than is necessary and the arterial pCO_2 falls, leading to a rise in pH. Respiratory alkalosis is most commonly a transient phenomenon due to hysteria when a person imagines that she is not breathing rapidly enough: she therefore tries to overbreathe and gets rid of too much carbon dioxide so producing alkalosis and tetany. More prolonged overbreathing may occur during artificial respiration, as a result of damage to the respiratory control centre by drugs such as aspirin, or as a result of an ascent to high altitude. In the last of these situations, oxygen lack drives ventilation at a rate greater than that necessary for the elimination of carbon dioxide in the usual way. If the hyperventilation persists, the alkalosis can be corrected by the kidney. Because of the low pCO_2 and the high pH, the secretion of hydrogen ions by the tubules falls. This means that more bicarbonate is filtered that can be reabsorbed. Bicarbonate is therefore lost and hydrogen ions are conserved. The process continues until the low pCO_2 is compensated for by a low bicarbonate concentration and pH is normal.

METABOLIC ACIDOSIS

This can occur when excessive amounts of hydrogen ions are produced as in diabetic coma or when the kidneys cease secreting hydrogen ions normally as in renal failure. Either way the concentration of hydrogen ions in the blood rises and the bicarbonate concentration falls. This time it is the respiratory system which attempts to compensate and to restore arterial pH to normal. The low pH stimulates the chemoreceptors in the aorta and carotid bodies. This leads to a rise in ventilation rate and to a fall in pCO_2 which tends to return blood pH to normal. Compensation is never complete because a sufficient increase in ventilation never occurs, as the stimulation because of the low pH is counteracted by inhibition because of the low pCO_2.

METABOLIC ALKALOSIS

This occurs when excessive amounts of hydrogen ions are lost from the body. By far the commonest cause is vomiting of acid gastric juice. Since the secretion of one hydrogen ion into the stomach is accompanied by the secretion of one bicarbonate ion into the blood, if the hydrogen ions are not absorbed further down the gut plasma bicarbonate concentration rises and the blood becomes more alkaline. The alkalinity depresses respiration by acting on the chemoreceptors and the resulting rise in the pCO_2 partly compensates for the rise in bicarbonate.

Index